Compression Textiles for Medical, Sports, and Allied Applications

Textile-based compression therapy is used in a range of applications, such as for athlete and sport recovery, enhanced proprioception, compression spacesuits, and in the management of chronic diseases. This book provides an overview of compression devices and products, testing methods to measure the properties of materials used in compression devices, and design considerations based on dynamic body measurements. It also includes a model for predicting pressure and details the challenges in applying compression for various applications.

Chapters in this book:

- Discuss the science behind compression therapy
- Delve into the materials used in compression devices and products and assesses their performance based on their properties and structure
- Cover theoretical modeling to predict the pressure exerted by compression devices on the human body
- Consider compression textile design based on dynamic body measurements

This book is aimed at professionals and researchers in textile engineering, materials engineering, biotechnology, and the development of textile-based compression devices and products, and at such medical practitioners as phlebologists.

Textile Institute Professional Publications

Series Editor: Helen D. Rowe, *The Textile Institute*

The Grammar of Pattern
Michael Hann

Standard Methods for Thermal Comfort Assessment of Clothing
Ivana Špelić, Alka Mihelić Bogdanić, and Anica Hursa Sajatovic

Fibres to Smart Textiles
Advances in Manufacturing, Technologies, and Applications
Asis Patnaik and Sweta Patnaik

Flame Retardants for Textile Materials
Asim Kumar Roy Choudhury

Textile Design
Products and Processes
Michael Hann

Science in Design
Solidifying Design with Science and Technology
Tarun Grover and Mugdha Thareja

Textiles and Their Use in Microbial Protection
Focus on COVID-19 and Other Viruses
Jiri Militky, Aravin Prince Periyasamy, and Mohanapriya Venkataraman

Dressings for Advanced Wound Care
Sharon Lam Po Tang

Medical Textiles
Holly Morris and Richard Murray

Odour in Textiles
Generation and Control
G. Thilagavathi and R. Rathinamoorthy

Principles of Textile Printing
Asim Kumar Roy Choudhury

Solar Textiles
The Flexible Solution for Solar Power
John Wilson and Robert Mather

Digital Fashion Innovations: Advances in Design, Simulation, and Industry
Abu Sadat Muhammad Sayem

Compression Textiles for Medical, Sports, and Allied Applications
Nimesh Kankariya and René M. Rossi

For more information about this series, please visit:
www.routledge.com/Textile-Institute-Professional-Publications/book-series/TIPP

Compression Textiles for Medical, Sports, and Allied Applications

Edited by

Nimesh Kankariya and René M. Rossi

CRC Press is an imprint of the
Taylor & Francis Group, an **informa** business

Designed cover image: Nimesh Kankariya

First edition published 2024
by CRC Press
6000 Broken Sound Parkway NW, Suite 300, Boca Raton, FL 33487-2742

and by CRC Press
4 Park Square, Milton Park, Abingdon, Oxon, OX14 4RN

CRC Press is an imprint of Taylor & Francis Group, LLC

Library of Congress Cataloging-in-Publication Data

Names: Kankariya, Nimesh, editor. | Rossi, René, editor.
Title: Compression textiles for medical, sports, and allied applications /
edited by Nimesh Kankariya and René Rossi.
Other titles: Textile Institute professional publications
Description: First edition. | Boca Raton : CRC Press, 2024. | Series:
Textile Institute professional publications | Includes bibliographical
references and index.
Identifiers: LCCN 2023002034 (print) | LCCN 2023002035 (ebook) | ISBN
9781032287942 (hbk) | ISBN 9781032287904 (pbk) | ISBN 9781003298526
(ebook)
Subjects: MESH: Compression Bandages | Vascular Diseases--therapy |
Intermittent Pneumatic Compression Devices | Textiles | Equipment Design
Classification: LCC RC691.6.B55 (print) | LCC RC691.6.B55 (ebook) | NLM
WG 26 | DDC 616.1/306--dc23/eng/20230508
LC record available at https://lccn.loc.gov/2023002034
LC ebook record available at https://lccn.loc.gov/2023002035

ISBN: 9781032287942 (hbk)
ISBN: 9781032287904 (pbk)
ISBN: 9781003298526 (ebk)

DOI: 10.1201/9781003298526

Typeset in Times
by Deanta Global Publishing Services, Chennai, India

Contents

Series Preface

TEXTILE INSTITUTE PROFESSIONAL PUBLICATIONS

The aim of the Textile Institute Professional Publications is to provide support to textile professionals in their work and to help emerging professionals, such as final year or Master's students, by providing the information needed to gain a sound understanding of key and emerging topics relating to textile, clothing and footwear technology, textile chemistry, materials science, and engineering. The books are written by experienced authors with expertise in the topic and all texts are independently reviewed by textile professionals or textile academics.

The textile industry has a history of being both an innovator and an early adopter of a wide variety of technologies. There are textile businesses of some kind operating in all counties across the world. At any one time, there is an enormous breadth of sophistication in how such companies might function. In some places where the industry serves only its own local market, design, development, and production may continue to be based on traditional techniques, but companies that aspire to operate globally find themselves in an intensely competitive environment, some driven by the need to appeal to followers of fast-moving fashion, others by demands for high performance and unprecedented levels of reliability. Textile professionals working within such organisations are subjected to a continued pressing need to introduce new materials and technologies, not only to improve production efficiency and reduce costs, but also to enhance the attractiveness and performance of their existing products and to bring new products into being. As a consequence, textile academics and professionals find themselves having to continuously improve their understanding of a wide range of new materials and emerging technologies to keep pace with competitors.

The Textile Institute was formed in 1910 to provide professional support to textile practitioners and academics undertaking research and teaching in the field of textiles. The Institute quickly established itself as the professional body for textiles worldwide and now has individual and corporate members in over 80 countries. The Institute works to provide sources of reliable and up-to-date information to support textile professionals through its research journals, the *Journal of the Textile Institute*[1] and *Textile Progress*,[2] definitive descriptions of textiles and their components through its online publication *Textile Terms and Definitions*,[3] and contextual treatments of important topics within the field of textiles in the form of self-contained books such as the Textile Institute Professional Publications.

REFERENCES

http://www.tandfonline.com/action/journalInformation?show=aimsScope&journalCode=tjti20

http://www.tandfonline.com/action/journalInformation?show=aimsScope&journalCode=ttpr20

http://www.ttandd.org

Foreword

When I was approached to write a 'Foreword' for the book entitled *Compression Textiles for Medical, Sports and Allied Applications*, I asked myself whether I am the right person to comment on the book and suggest readers accordingly. Eventually, I accepted the invitation mainly for three reasons : (i) the title and the contents of the book chapters are one of my favourite subject areas within Medical Textiles; (ii) the academic experience that I gained in specialised textile-based medical devices over 40 years, which incurred many publications (184), patents (8) and books (11); and (iii) the prestigious medal that I received from the Textile Institute, Manchester, for my devotion to the furtherance of scientific knowledge concerning fibrous materials.

Medical intervention utilising hi-tech textile medical devices has been crucial in various treatment protocols. It has long been established that compression therapy, considered the gold standard treatment, by making use of compression bandages is an effective treatment for preventing, treating and managing chronic venous insufficiency and leg ulceration despite surgical strategies, electromagnetic therapy and Intermittent Pneumatic Compression (IPC). The other application of compression textiles, for instance treating and managing Deep Vein Thrombosis (DVT), varicose veins, burn scars, muscle injuries and orthopaedic disorders, is also gaining importance. In these circumstances, I believe, this book provides essential updates to the readers in various aspects of compression textiles. It takes a holistic look at all aspects of compression textiles and comprehensively reviews the achievements till date. Some of the contents of the chapters such as 2D, 3D and 4D body scanning techniques, pressure sensors used for appropriate product development and application of compression garments in sports and space attract my special interest, and no doubt these would attract the readers as well.

The editors have extensive experience in these specialised subject areas and hold impressive academic outputs to their credits. They have carefully designed various subject areas within compression textiles and have taken excellent initiative to edit this book. The contributors of the 12 chapters are eminent academics, and their chapters are commended for the depth and breadth of the work described. These chapters address the current and timely updates in compression textiles.

I firmly believe that this book will prove to be an inspiration to readers and will trigger new thoughts and developmental ideas in compression textiles. I am certain that it will make significant contributions to the understanding of various issues related to the development of textile-based compression medical devices and be of great use to peers in academia, textile scientists, medical textiles companies and medical professionals.

Professor S. Rajendran PhD AIC FICS CText FTI
Emeritus Professor
Formerly Professor of Biomedical Materials & STEM Champion
The University of Bolton
Bolton, United Kingdom

Editors

Nimesh Kankariya serves as Director for Textiles and Materials Research Limited, New Zealand (TexMat Research). Much of his recent research focuses on smart textiles, including compression textiles and emerging technologies. He has been involved in several industrial-based research projects. Nimesh's doctoral thesis "Textiles and Compression of the Lower Limb" is on the Sciences Divisional list of Exceptional Doctoral Theses, University of Otago, New Zealand.

René M. Rossi has worked at Empa, the Swiss Federal Laboratories for Materials Science and Technology, for more than 25 years in the field of the interactions between materials and the human skin. He studied physics and obtained his PhD at ETH Zurich, Switzerland. Since 2003, Prof. Rossi has been leading the Laboratory for Biomimetic Membranes and Textiles at Empa, a group of around 40 researchers developing novel smart fibers, textiles, and membranes for body monitoring, drug delivery, and tissue engineering applications. He is Adjunct Professor at the Department of Health Sciences and Technology at ETH Zurich and Invited Professor at the University of Haute-Alsace in Mulhouse, France.

Contributors

Brubacher, Kristina (Dr.)
Department of Materials
The University of Manchester
Manchester, United Kingdom

Bye, Elizabeth (Dr.)
College of Design
University of Minnesota
St. Paul, MN, USA

Camenzind, Martin
Empa, Lab for Biomimetic Membranes and Textiles
St. Gallen, Switzerland

Chattopadhyay, Rabisankar (Dr.)
Department of Textile and Fiber Engineering
Indian Institute of Technology Delhi
New Delhi, India

Clarke, Jonathan
Human Aerospace Pty Ltd.
Melbourne, Australia

and

Mars Society Australia
Melbourne, Australia

Cloet, Alison
College of Design
University of Minnesota
St. Paul, MN, USA

Griffin, Linsey (Dr.)
College of Design
University of Minnesota
St. Paul, MN, USA

Kankariya, Nimesh (Dr.)
Textiles and Materials Research Limited
New Zealand

and

University of Otago
Dunedin, New Zealand

Khairi, Mohd Yusoh
Negeri Sembilan Police Contingent
Royal Malaysia Police, Malaysia

Kyosev, Yordan (Dr.)
Institut für Textilmaschinen und Textile, Hochleistungswerkstofftechnik (ITM)
Technische Universität Dresden
Dresden, Germany

MacRae, Braid A. (Dr.)
Human Aerospace Pty Ltd.
Melbourne, Australia

and

RMIT University
Melbourne, Australia

Rossi, René M. (Dr.)
Empa, Lab for Biomimetic Membranes and Textiles
St. Gallen, Switzerland

Rudakov, Abby
School of Fashion and Textiles
RMIT University
Melbourne, Australia

Ruznan, Wan Syazehan (Dr.)
Department of School of Industrial Technology
Faculty of Applied Sciences
Universiti Teknologi MARA
Negeri Sembilan Branch (Kuala Pilah Campus)
Shah Alam, Malaysia

Spahiu, Tatjana (Dr.)
Mechanical Engineering Faculty
Polytechnic University of Tirana
Tirana, Albania

Stämpfli, Rolf
Empa, Lab for Biomimetic Membranes and Textiles
St. Gallen, Switzerland

Tohid, Nurfazieyana Ahmad
Department of School of Industrial Technology
Faculty of Applied Sciences
Universiti Teknologi MARA
Shah Alam, Selangor, Malaysia

Waldie, James
Human Aerospace Pty Ltd.
Melbourne, Australia

and

RMIT University
Melbourne, Australia

Wang, Yongrong (Dr.)
Fashion and Design College
Donghua University
Shanghai, China

Yusof, Nur Ain (Dr.)
Department of School of Industrial Technology
Faculty of Applied Sciences
Universiti Teknologi MARA
Shah Alam, Selangor, Malaysia

1 Overview of Compression Therapy

Nimesh Kankariya

CONTENTS

1.1 HISTORICAL DEVELOPMENTS IN COMPRESSION THERAPY

The history of applying pressure on the lower leg is lost in the mists of time until the oldest evidence is discovered in the paintings on the rocks of Tassili caves in the Sahara which represented the Neolithic period (5100–2500 BC) (Partsch, Rabe and Stemmer 1999). These paintings depicted the compression by means of bandages which were wrapped around the legs of the warriors while performing a kind of ritual dance. Additional testimonies to use compression wraps to heal wounds and ulcers are found in ancient periods of the Greeks, Egyptians, Romans, Hebrews, and Hippocrates; however, in these periods the way to apply compression on the legs may differ compared to the way which today is known as compression therapy (Orbach 1979, Partsch, Rabe and Stemmer 1999). The development stages of compression therapy include identification of the relationship between compression therapy and blood flow, application of gradient pressure from proximal to distal of the lower leg, and discovery and invention of different types of compression products, i.e. bandages, stocking, and pneumatic compression. A chronology of key milestones in the development of compression therapy is shown in Table 1.1.

DOI: 10.1201/9781003298526-1

TABLE 1.1
Historical Developments in Compression Therapy

Timeline	Development descriptions	References
51st–26th century BC	Images of the warriors (in Neolithic era) were found with bandaged lower leg in the painting of Tassili caves in the Sahara.	Partsch, Rabe and Stemmer (1999)
17th–16th century BC	Evidence of the use of compression wrap on the legs was found in the Egyptian's 17th dynasty (1650–1552 BC), described in the Edwin Smith Papyrus (155 BC).	Lippi, Favaloro and Cervellin (2011)
8th century BC	Hebrews, Greeks, and Romans used bandages to heal wounds and ulcers, mentioned in the book Isaiah (verse 6 of chapter 1).	Orbach (1979), Partsch, Rabe and Stemmer (1999)
5th–2nd century BC	"Corpus Hippocraticum", a medicine text written by Hippocrates (460–373 BC), and an Indian medicine text "Sushruta Samhita" (200 BC) reported the use of *linen bandages*.	Orbach (1979), Lippi, Favaloro and Cervellin (2011)
1st century BC–2nd century AD	Aurelius Cornelius Celsus (25 BC–50 AD) and Galen (130 AD–200 AD) reported the use of wool-linen compression bandages to prevent blood reflux.	Orbach (1979), Lippi, Favaloro and Cervellin (2011)
14th–16th century AD	Wrapping the bandage from foot to knee was recommended by Giovanni Michele Savonarola (1384–1468 AD), who is considered as the founder of the conventional treatment of varicose veins according to his text "Practice". Inelastic materials including dog leather were the key components in the compression bandages.	Lippi, Favaloro and Cervellin (2011), Partsch (2017)
17th century AD	William Harvey (1578–1657 AD) explained blood circulation and revealed the relationship between compression therapy and venous stasis in 1628 AD. A laced stocking composed of dog leather was developed by Richard Wiseman (1622–1676) and used on legs to heal ulcers.	Orbach (1979), Lippi, Favaloro and Cervellin (2011)
19th century AD	Elastic stockings made up of natural rubber and India rubber appeared at the start of the 19th century AD. Although the rubber materials used were not durable, stockings were poorly compliant for the users, and stockings were not easy to put on and take off.	The compression study group (2009)
	1839 – The discovery of vulcanization process to make rubber durable by forming the cross-links between polymer chains by Charles Goodyear in 1839 created a breakthrough point in the textile industry.	Goodyear (1844)

(*Continued*)

TABLE 1.1 CONTINUED
Historical Developments in Compression Therapy

Timeline	Development descriptions	References
	1846 – Fine rubber thread with a quadrangular cross-section was developed and patented.	William and Thomas (1846)
	1848 – **Elastic stocking** weaved on the handloom with India rubber was developed and patented. This achievement is considered the birth of the modern elastic stocking.	William (1848)
	1885 – Zinc paste boot (first industrial adhesive bandage) introduced by Paul Unna in 1885 was marked as another historical point. Later on his pupil, Heinrich Fischer, in 1910, suggested applying the "**Unna boot**" for treating venous thrombosis.	Partsch (2017)
20th century AD	**1914** – First **seamless stocking** was fabricated and patented.	Scott (1914)
	The inventions of **Nylon** fiber in 1938, and **Spandex** in 1958 ushered in a compression textile revolution.	Wallace and Wilmington (1938), Shivers (1958)
	1951 – First pneumatic compression device was developed.	Brown (1951)
	1964 – Tubular bandage was developed.	Scholl Manufacturing Co Inc (1964)
	1980 – First four-layer compression bandage was developed by Charing Cross Hospital, London.	Moffatt (2004)
21st century AD	Multiple components, multi-layer bandaging, and stocking kits with improved thermophysiological and sensorial comfort.	Kankariya (2022)
	Advanced multi-section pneumatic compression devices equipped with digital programming.	
	Stimuli-responsive compression products with the capability of producing controlled compression on the lower leg.	

1.2 COMPRESSION THERAPY: ACTION AND MODALITIES

Compression therapy is a recognized method of choice for the treatment of a range of venous diseases, i.e. varicose veins, deep vein thrombosis, venous eczema, edema, and ulcers. It is also used for other applications such as sports and athletic recovery and enhances extravehicular activity (Waldie et al. 2002, Fletcher et al. 2013, Zamporri and Aguinaldo 2018, Franke, Backx, and Huisstede 2021). The compression device applies consistent pressure to the lower leg, causing the vein width to narrow, and the pressure in the veins of the limb is increased. This increased pressure allows more blood to flow towards the heart, reducing the risk of blood clot formation (Dissemond et al. 2016). The increased pressure in the veins also stimulates pumping of the lymphatic system, encourages re-absorption of lymphatic fluid, if any, and reduces fluid leakage out of the capillaries into interstitial space. These combined actions (more reabsorption and less leaking of fluid) help manage and/or reduce the edema and prevent the development of leg ulcers (Lymphoedema Framework 2006, Lim and Davies 2014). In the lying position, low pressure (~15 mmHg) is sufficient to constrict the veins followed by an acceleration in blood flow. If the same effect is to be accomplished in a standing position, higher pressure (60–90 mmHg) is required (Dissemond et al. 2016).

Three major compression product categories available are compression bandages, stockings, and pneumatic compression devices (Partsch and Mortimer 2015). A description of compression products and their further classification is given in the following sections. Adoption of compression modalities is dependent on the specific applications, i.e. the severity of venous conditions and the patient's tolerance of the device and associated perception of comfort (Andre and Pierre 2014, Nair 2014).

1.2.1 Bandages

Bandages are long strips of fabric that may be wrapped around the body forming a continuous covering which applies an external force (Fletcher et al. 2013, Dolibog et al. 2014). Bandage strips can be knitted, woven, or non-woven and can be designed using elastic or inelastic materials (Rajendran, Anand and Rigby 2016). These bandage fabrics can also have different yarn structures such as core spun, filament yarns, and different fiber compositions such as nylon, cotton, wool, Lycra®, rubber, and non-latex materials (Kankariya 2022, Kumar, Das and Alagirusamy 2014b, Sikka, Ghosh and Mukhopadhyay 2014). Bandages are classified in a number of ways based on their extensibility, elasticity, number of layers, and material performance (Thomas 1997, Partsch 2008) as per the following sub-sections.

1.2.1.1 Based on Material Performance

Compression bandages designed to exert a specified sub-bandage pressure to a fixed ankle circumference of 230 mm with an overlap of 50% between successive layers can be classified into four groups: light, moderate, high, and extra high (Thomas 1990, Nelson 1995, Thomas 1997, Mulder 2002, Williams 2002, Clark 2003). Different countries have their own standards to define the pressure range of compression bandages. The two most common standards are the British standard (BS 7505: 1995) (light: <20, moderate: 21–30, high: 31–40, extra high: 41–60) and the German standard (RAL-GZ 387: 1987) (light: 18.4–21.2, moderate: 25.1–32.1, high: 36.4–46.5, extra high: >59) (British Standards Institution 1995, German Institute for Quality Assurance and Labeling 2008).

1.2.1.2 Based on Elasticity

The bandage can be elastic and inelastic in nature. Elastic materials return to their original dimension once the extension forces are removed (Morton and Hearle 2008, Partsch 2008). The elasticity of compression systems has been further characterized by stiffness which influences the performance of a bandage system. The European pre-standard for medical compression fabric refers to stiffness as the increase in pressure per 10 mm increase in leg circumference (Partsch 2005, Mosti and Mattaliano 2007). The bandage, combined with different layers of elastic material applied over each other, converts into a stiff device because of the friction between the layers, i.e. the four-layer bandage, although having elastic components in each layer performs as a stiff bandage (Mosti, Mattaliano and Partsch 2008, Hirai, Koyama, Miyazaki, Iwata et al. 2012).

1.2.1.3 Based on Extensibility

The extensibility of a bandage refers to the increase in length that occurs in response to applied stress. The physical structure of a bandage is such that it restricts further stretching once a certain degree of extension is attained and is referred to as "a lock out" bandage (Morton and Hearle 2008, Partsch 2008). Inelastic bandages, i.e. zinc oxide bandages, neither extend nor return to their original shape with external force. When applied without stretch, the resting pressure of these bandages remains low. Inelastic short-stretch bandages ideally lock out at approximately 30% stretch (70% extension). These bandages produce high pressure during standing and walking by resisting the increase of muscle volume. Medium stretch bandages have the ability to extend between 70% and 140%. Elastic long-stretch bandage only lock out with extensibility of more than 140%. Due to their elastic property, long-stretch bandages tend to return to the original shape and exert high working and resting pressure. Users can find the sustained pressure of the bandage on limbs intolerable at night due to high resting pressure (Mulder 2002, Kumar, Das and Alagirusamy 2014a).

1.2.1.4 Based on Layers and Components

The compression bandage can be single-layer, or multi-layer, and the bandage material can be made up of single and multiple materials (single component, multiple components) (Gupta, Koven, Lester Shear et al. 2000, Vowden, Mason, Wilkinson and Vowden 2000, Dale et al. 2004, Jünger, Partsch, Ramelet and Zuccarelli 2004, Mosti, Mattaliano and Partsch 2008, Dabiri, Hammerman, Carson and Falanga 2015). Multi-layer bandages could be formed by more than two layers of a single material or, in the case of multi-layer bandages such as the four-layer bandage systems, by multiple layers of different bandage materials (Partsch 2008). There is a growing market for use of both multi-component kits and multi-layer bandages. In multi-component bandages the components are of different materials which have different effects on the sub-bandage pressure functions such as protection, retention, or padding (Partsch 2008, Kumar, Das and Alagirusamy 2014a).

1.2.1.5 Adhesive and Cohesive Bandage

The conical shape of the human leg may necessitate the use of adhesive and/or cohesive compression bandages to prevent the slippage of the bandage over the lower leg. The cohesive bandage or self-adhering bandage is a type of bandage or wrap that adheres only to itself and not to the skin or hair, whereas the adhesive bandage, which is sticky on one side, adheres to skin and textile surfaces. Adhesive and cohesive bandages are made of elasticated textile fibers/yarns with either woven or knitting technologies. The textile fabrics of adhesive/cohesive bandages are coated with an adherent ingredient, such as polyacryl or zinc oxide, that allows the bandages to stick to a variety of surfaces, such as fabric surface and skin surface. Both cohesive and adhesive bandages provide a higher level of compression compared to plain or crepe bandages. In addition, when applied correctly, cohesive and adhesive bandages do not loosen over time resulting in a more consistent, long-term level of compression. However, the adhesive bandage is non-washable and can only be used once (Campbell 2004, Charles, Moore and Varrow 2003, Vowden, Vowden and Partsch et al. 2011, Hafner, Botonakis and Burg 2000).

1.2.2 Stockings

Compression stockings made from elasticated knitted fabric are widely used for pressure therapy as they produce graduated pressure over the lower extremities. Graduated compression stockings exert the highest pressure at the ankle, with the pressure gradually diminishing as the limb circumference increases towards the calf (Horner, Fernandes, Fernandes and Nicolaides 1980, Burnand and Layer 1986, Liu, Guo, Lao and Little 2016). This graduated pressure assists blood movement upwards towards the heart resisting backward movement and accumulation in the veins (Lim and Davies 2014). The applied pressure depends on the structure of the knit and fit of the stocking. The size of the stocking, by design, is smaller than the size of the limb they enclose, ensuring stretch during use. This form of the product is less bulky and more comfortable than other compression devices, i.e. inelastic bandages and elastic bandages. Stockings are mainly used to prevent edema and DVT generation and the recurrence of ulcer after healing (Ramelet 2002, Moneta and Partsch Accessed: 31/3/2017). Compression stockings are sold in different sizes and lengths, thigh length or knee length (Loomba, Arora, Chandrasekar and Shah 2012).

Compression stockings are classified, Class I – low pressure to Class IV – high pressure, based on the pressure they exert on the lower extremity (Kankariya, Laing, Wilson 2021b). They are also characterized by properties such as stiffness (Section 1.2.1) and hysteresis, the resistance of materials due to internal friction (Neumann et al. 2016). Hysteresis of elastic compression stocking refers to the loss in recovered linear length after exposure to cyclically repeated stretch-relaxation during walking, i.e. as the circumference of the leg changes, requiring the materials of the stocking to adjust to these changes (Neumann 2013, Neumann et al. 2016). Hysteresis is affected by walking speed. During normal walking, elastic material with high hysteresis will be unable to keep synchronized with these circumferential changes.

1.2.3 Pneumatic Compression Devices

Pneumatic compression is a concept that has been used to treat venous disease in the lower limb for over 170 years (Kumar and Alexander 2002). Pneumatic compression systems consist of an inflatable sleeve composed of either single or multiple pressure chambers that encircle the lower leg. The pressure chambers of the pneumatic device are typically fabricated by air-impervious materials (e.g. polyurethane, polyethylene, silicone), and a next-to-skin layer is generally used under these pressure chambers (Kankariya 2020, Kankariya, Laing, Wilson 2020, Kankariya, Wilson and Laing 2021, Kankariya, Laing and Wilson 2021a). An electrical or mechanical pneumatic pump connected to the sleeve fills chambers intermittently with compressed air. In a clinical context, the sleeve inflates and deflates to generate the desired pressure on the lower limb. The inflation and deflation time, total treatment time, and pressure exerted cyclically in treatment using intermittent compression devices vary (Rithalia, Heath and Gonsalkorale 2002). The effective pressure applied by the pneumatic device may be negotiated with the comfort compliance of the patients (Morris 2008).

Pneumatic compression devices, in contrast to conventional compression devices which generate a static force on the lower extremities and rely on the changes in muscle tone caused by the physical movement of the patients, simulate the muscle tone irrespective of the physical movement of the patients. These devices are helpful for use by people with obesity, or severely impaired movement for whom other options may be limited (Partsch 2008). The basis of using intermittent pneumatic compression is to aid the management of venous diseases; i.e. reduction of swelling and prevention of the recurrence of edema lie in the mechanical combination of the blood circulation effects (Chen, Frangos, Kilaru and Sumpio 2001). The mechanical effects of pneumatic devices relate to dynamic blood flow, structural technicalities of the blood vessels, surrounding skin tissues, design of the pressure device, cycle duration, application/use time, and pressure level. Blood dynamics affect the rate of supply of necessary nutrients and improvement in oxygen supply (Sayegh 1987). The consequences of using pneumatic devices may be summarized as increased blood flow, reduced swelling, enhanced vasodilation, and increased oxygen and other nutrients, improving the healing of ulcers. The pneumatic compression can be classified as given in the following sub-section.

1.2.3.1 Based on Complexity of Device

The first-generation intermittent devices, with a single inflatable pressure chamber, produce uniform and sustained pressure to the entire leg. Single chamber devices were non-programmable and not able to produce a pressure gradient (Feldman, Stout, Wanchai, Stewart et al. 2012). Manual control over pressure distribution is also not available in these devices. Therefore, single chamber devices are not usually recommended for the management of venous disorders (Partsch and Mortimer 2015).

Multi-chamber pneumatic devices were designed in the 1970s. These devices have three or four chambers connected with individual inflated sleeves. Each sleeve is connected to an individual pneumatic pump. Each chamber inflates sequentially from the distal part of the leg to the proximal part until all chambers are inflated. Subsequent to the inflation phase, all chambers are deflated simultaneously, and the cycle is repeated. These devices do not have any pressure programming options and also typically do not have any independent adjustability. These devices are able to produce a pressure gradient as higher pressure is applied in the distal section compared to lower in the proximal section (Feldman et al. 2012).

Multi-segmented devices with calibrated gradient compression have at least three zones of pressure. These devices also have a manual programming option and have independent adjustment options for the level and location of pressure. These devices have multiple chambers ranging from a minimum of 4 to up to 36 (Feldman et al. 2012).

The advanced multi-section pneumatic compression devices are equipped with digital programming. At one time, 1–2.5 chambers are activated as the pressure progresses in the distal to proximal direction. These devices actively produce cyclical pressure waves from resting to working pressure to simulate lymphatic drainage techniques, and as a consequence work to increase the fluid velocity (Partsch 2008, Feldman et al. 2012). Partsch concluded that compression therapy using an intermittent pneumatic device did not only apply the required level of pressure with a correct gradient to the leg but also simulated rhythmic muscle contractions even when such active movements were either restricted or absent (Partsch 2008).

1.2.3.2 Based on Inflatable Sleeves Encircled around the Lower Leg

Circumferential pneumatic devices encircle and compress the whole leg and are often configured either with a zipper or are often configured either without a size adjustment mechanism such as a zipper or with Velcro® or another adhesive which allows for size adjustment (Dai, Gertler and Kamm 1999, Morris and Woodcock 2004). Air pumps are used to fill the garment and apply pressure (Morris 2008).

Non-circumferential bladders encircle only part of the leg circumference, resulting in less air is required to inflate the bladders. The non-circumferential pneumatic device is applied on the posterior part of the lower leg as deep veins which are responsible for flowing 90% of the blood back to the heart with the help of a muscle pump are located there (Youn and Lee 2019). The material over the part of the leg without the bladder must not stretch for the non-circumferential wrap-around devices. Pressure is thus applied as a result of inflation of the posterior bladder. If the garment is free to slide over the skin surface, then the pressure profile obtained by the non-circumferential designs should match that of circumferential devices (Morris 2008).

1.3 CONTRADICTIONS AND CHALLENGES TO COMPRESSION THERAPY

Research studies concluded that applying some compression is preferable to no compression at all (Mosti and Partsch 2013, Carvalho, Lopes, Guerreiro and Pereira 2015, Mosti et al. 2015). However, there is presently no compelling data to suggest which compression modality is better successful for a given application. Because the mechanism(s) of action of compression therapy is not well understood, there is debate on the appropriate degree of pressure required for various applications to achieve significant improvements. At the time of publication, different national pressure classification systems of compression devices are discrepant. A project to develop a unified pressure classification system for compression garments is presently ongoing throughout Europe. However, Asia, Africa, America, and Oceania regions currently do not use a defined standard pressure classification system. In addition, even within the same country, products from different manufacturers differ. The pressure exerted by medical compression garments is measured using a variety of pressure measurement instruments (i.e. HATRA pressure tester, HOSY pressure tester, Sigvaris interface-pressure gauge tester, Borgnis medical stocking tester, PicoPress®). Because each instrument is distinct in terms of design, methodology, and whether it uses a direct (on-body) or indirect (off-body) technique, the pressures measured for the same garment vary (Meghan Hegarty-Craver et al. 2015). Therefore, an end user may receive class II compression stockings with a pressure between 18 and 28 mmHg from manufacturer X and between 28 and 38 mmHg from manufacturer Y. Because of the complex supply chains of manufacturers and distributors of medical textile products, as well as the relatively open markets within countries, a wide range of compression products is potentially accessible in any one country. Differences among countries in product sizing, product size designations, and specification for use are ongoing international problems present in medical and non-medical sectors.

1.4 MARKET TRENDS IN COMPRESSION PRODUCTS

The global compression textiles market is expected to rise to US$3.9 billion by 2026, up from US$3.0 billion in 2021, at a Compound Annual Growth Rate (CAGR) of 5.2% (Market and Market Research Pvt Ltd Accessed on 7 Feb 2021). During the projected period, emerging economies such as India, China, Mexico, and Brazil are expected to provide considerable potential possibilities for compression therapy market competitors. The market for compression therapy is projected to be dominated by compression garments (Market and Market Research Pvt Ltd Accessed on 7 Feb 2021).

The rising target patient populations (i.e. patients with ulcer, edema, varicose veins, obesity (excessive weight gain), diabetes, lymphedema, burn scar injury) around the world, prolonged stagnation of the human body, the rise in sports, athletics, and fitness activities, and growing awareness among practitioners and patients about the latest treatment options available for various orthopedic and vascular disorders are the major factors driving the growth of this market. However, poor patient compliance with textile-based compression devices, disparities in pressure classifications of compression products between nations, and the lack of internationally agreed pressure measures are the key factors that are projected to restrain the growth of this market throughout the forecast period (Market and Market Research Pvt Ltd Accessed on 7 Feb 2021).

REFERENCES

Andre, C T, and Pierre, B. 2014. "Chronic venous disease during pregnancy." *Phlebolymphology* 21 (3):138–145.

British Standards Institution. 1995. "BS 7505." In *Specification for the elastic properties of flat, non-adhesive, extensible fabric bandages*. London: British Standards Institute.

Brown, W J. 1951. "Pneumatic pressure garment." Patent no. US 2694395A. Filed 10/5/1951, and Issued 16/11/1954.

Burnand, K G, and Layer, G T. 1986. "Graduated elastic stockings." *British Medical Journal* 293 (6541):224–225.

Campbell, L. 2004. "Effectiveness in retention bandaging." *Practice Nursing* 15 (2):97–98.

Carvalho, C A, Lopes, P R, Guerreiro, G, and Pereira, G J M. 2015. "Reduction of pain and edema of the legs by walking wearing elastic stockings." *International Journal of Vascular Medicine* 2015:1–4. doi:10.1155/2015/648074.

Charles, H, Moore, C, and Varrow, S. 2003. "Cohesive short stretch bandages in the treatment of venous leg ulceration." *British Journal of Community Nursing* 8 (3 Suppl):17–22.

Chen, A H, Frangos, S G, Kilaru, S, and Sumpio, B E. 2001. "Intermittent pneumatic compression devices – Physiological mechanisms of action." *European Journal of Vascular and Endovascular Surgery* 21 (5):383–392. doi:10.1053/ejvs.2001.1348.

Clark, M. 2003. "Compression bandages: Principles and definitions." In *Understanding compression therapy*, edited by EWMA Position document. London.

The Compression Study Group. 2009. *Compression – Consensus document based on scientific evidence and clinical experiences.* Edited by Fabrizio M. Torino. Italy: The Compression Study Group.

Dabiri, G, Hammerman, S, Carson, P, and Falanga, V. 2015. "Low-grade elastic compression regimen for venous leg ulcers – An effective compromise for patients requiring daily dressing changes." *International Wound Journal* 12 (6):655–661. doi:10.1111/iwj.12186.

Dai, G, Gertler, J P, and Kamm, R D. 1999. "The effects of external compression on venous blood flow and tissue deformation in the lower leg." *Journal of Biomechanical Engineering* 121 (6):557–564. doi:10.1115/1.2800853.

Dale, J J, Ruckley, C V, Gibson, B, Brown, D, Lee, A J, and Prescott, R J. 2004. "Multilayer compression: Comparison of four different four-layer bandage systems applied to the leg." *European Journal of Vascular and Endovascular Surgery* 27 (1):94–99. doi:10.1016/j.ejvs.2003.10.014.

Dissemond, J, Assenheimer, B, Bültemann, A, Gerber, V, Gretener, S, Kohler-von Siebenthal, E, Koller, S, Kröger, K, Kurz, P, Läuchli, S, Münter, C, Panfil, E M, Probst, S, Protz, K, Riepe, G, Strohal, R, Traber, J, and Partsch, H. 2016. Compression therapy in patients with venous leg ulcers. *Journal der Deutschen Dermatologischen Gesellschaft* 14 (11):1072–1087. doi:10.1111/ddg.13091. PMID: 27879096.

Dolibog, P, Franek, A, Taradaj, J, Dolibog, P, Blaszczak, E, Polak, A, Brzezinska-Wcislo, L, Hrycek, A, Urbanek, T, Ziaja, J, and Kolanko, M. 2014. "A comparative clinical study on five types of compression therapy in patients with venous leg ulcers." *International Journal of Medical Sciences* 11 (1):34–43. doi:10.7150/ijms.7548.

Feldman, J L, Stout, N L, Wanchai, A, Stewart, B R, Cormier, J N, and Armer, J M. 2012. "Intermittent pneumatic compression therapy: A systematic review." *Lymphology* 45 (1):13–25.

Fletcher, J, Moffatt, C, Partsch, H, Vowden, K, and Vowden, P. 2013. Principles of compression in venous disease: A practitioner's guide to treatment and prevention of venous leg ulcers. *Wounds International* 1 (2013):1–21.

Franke, T, Backx, F, and Huisstede, B. 2021. Lower extremity compression garments use by athletes: Why, how often, and perceived benefit. *BMC Sports Science Medicine and Rehabilitation* 31 (2021):1–14.

German Institute for Quality Assurance and Labeling. 2008. "RAL-GZ 387/1." In *Medical compression hosiery – Quality assurance.* Berlin: RAL, German Institute for Quality Assurance and Labeling.

Goodyear, C. 1844. "Improvement in india-rubber fabrics." Patent no. 3633. Issued 15/6/1844.

Gupta, A K, Koven, J D, Lester, R, Shear, N H, and Sauder, D N. 2000. "Open-label study to evaluate the healing rate and safety of the profore™ extra four-layer bandage system in patients with venous leg ulceration." *Journal of Cutaneous Medicine and Surgery* 4 (1):8–11. doi:10.1177/120347540000400103.

Hafner, J, Botonakis, I, and Burg, G. 2000. A comparison of multilayer bandage systems during rest, exercise, and over 2 days of wear time. *Archives of Dermatology* 136 (7):857–863.

Hirai, M, Koyama, A, Miyazaki, K, Iwata, H, and Kominami, Y. 2012. "Interface pressure and stiffness in different combinations of compression material." *Phlebology* 27 (2):82–89. doi:10.1258/phleb.2011.011010.

Horner, J, Fernandes, J, Fernandes, E, and Nicolaides, A N. 1980. "Value of graduated compression stockings in deep venous insufficiency." *British Medical Journal* 280 (6217):820–821. doi:10.1136/bmj.280.6217.820.

Jünger, M, Partsch, H, Ramelet, A A, and Zuccarelli, F. 2004. "Efficacy of a ready-made tubular compression device versus short-stretch compression bandages in the treatment of venous leg ulcer." *Wounds* 16 (10):313–320.

Kankariya, N. 2020. "Characterising a nonwoven component of a textile based compression intervention." AICTE International Conference on "Recent trends in textiles", India.

Kankariya, N. 2022. "Material, structure, and design of textile-based compression devices for managing chronic edema." *Journal of Industrial Textiles* 52 (2022):1–35. doi:10.1177/15280837221118844.

Kankariya, N, Laing, R M, and Wilson, C A. 2020. "Challenges in characterising wool knit fabric component of a textile based compression intervention." International Virtuwool Research Conference, New Zealand: AgResearch.

Kankariya, N, Laing, R M, and Wilson, C A. 2021a. "Prediction of applied pressure on model lower limb exerted by an air pneumatic device." *Medical Engineering & Physics* 97 (2021):77–87. doi:10.1016/j.medengphy.2021.07.007.

Kankariya, N, Laing, R M, and Wilson, C A. 2021b. "Textile-based compression therapy in managing chronic oedema: Complex interactions." *Phlebology* 36 (2):100–113. doi:10.1177/0268355520947291.

Kankariya, N, Wilson, C A, and Laing, R M. 2021. "Thermal and moisture behavior of a multi-layered assembly for use in a pneumatic compression device." *Textile Research Journal* 92 (15–16):2669–2684. doi:10.1177/00405175211006942.

Kumar, B, Das, A, and Alagirusamy, R. 2014a. "Characterization of compression bandage." In *Science of compression bandages*, edited by Kumar B, Das A, Alagirusamy R. India: Woodhead Publishing, pp. 21–40.

Kumar, B, Das, A, and Alagirusamy, R. 2014b. "Effect of material and structure of compression bandage on interface pressure variation over time." *Phlebology* 29 (6):376–385. doi:10.1177/0268355513481772.

Kumar, S, and Alexander, W M. 2002. "The effects of intermittent pneumatic compression on the arterial and venous system of the lower limb: A review." *Journal of Tissue Viability* 12 (2):58–66. doi:10.1016/S0965-206X(02)80015-8.

Lim, C S, and Davies, A H. 2014. "Graduated compression stockings." *Canadian Medical Association Journal* 186 (10):E391–E398. doi:10.1503/cmaj.131281.

Lippi, G, Favaloro, E J, and Cervellin, G. 2011. "Prevention of venous thromboembolism: Focus on mechanical prophylaxis." *Seminars in Thrombosis and Hemostasis* 37 (3):237–251. doi:10.1055/s-0031-1273088.

Liu, R, Guo, X, Lao, T T, and Little, T. 2016. "A critical review on compression textiles for compression therapy: Textile-based compression interventions for chronic venous insufficiency." *Textile Research Journal* 87 (9):1121–1141. doi:10.1177/0040517516646041.

Loomba, R S, Arora, R R, Chandrasekar, S, and Shah, P H. 2012. "Thigh-length versus knee-length compression stockings for deep vein thrombosis prophylaxis in the inpatient setting." *Blood Coagulation & Fibrinolysis* 23 (2):168–171. doi:10.1097/MBC.0b013e32834cb25b.

Lymphoedema Framework. 2006. *Best practice for the management of lymphoedema.* International Consensus. London: MEP Ltd.

Market and Market Research Pvt Ltd. *Compression therapy market.* Accessed on 7 Feb 2021 https://www.marketsandmarkets.com/Market-Reports/compression-therapy-market-146548022.html

Meghan, H C, Cassandra, K, William, O, Edward, G, and Lawrence, R. 2015. "Towards characterizing the pressure profiles of medical compression hosiery: An investigation of current measurement devices and techniques." *The Journal of The Textile Institute* 106 (7):757–767. doi:10.1080/00405000.2014.941535.

Moffatt, C. 2004. "Four-layer bandaging: From concept to practice. Part 1: The development of the four-layer system." *World Wide Wounds.* (worldwidewounds.com).

Moneta, G L, and Partsch, H. "Compression therapy for venous disorders and venous ulceration." Accessed on 31 March 2017. http://www.veinforum.org/patients/vein-handbook/chapter-13-compression-therapy-for-venous-disorders-and-venous-ulceration.html.

Morris, R J. 2008. "Intermittent pneumatic compression—Systems and applications." *Journal of Medical Engineering & Technology* 32 (3):179–188. doi:10.1080/03091900601015147.

Morris, R J, and Woodcock, J P. 2004. "Evidence based compression: Prevention of stasis and deep vein thrombosis." *Annals of Surgery* 239 (2):162–171. doi:10.1097/01.sla.0000109149.77194.6c.

Morton, W E, and Hearle, J W S. 2008. "Tensile properties." In *Physical properties of textile fibres* (Fourth edition), edited by Morton, WE, and Hearle, JWS, 274–321. England: Woodhead Publishing. doi:10.1533/9781845694425.274.

Mosti, G B, and Mattaliano, V. 2007. "Simultaneous changes of leg circumference and interface pressure under different compression bandages." *European Journal of Vascular and Endovascular Surgery* 33 (4):476–482. doi:10.1016/j.ejvs.2006.11.035.

Mosti, G, Cavezzi, A, Partsch, H, Urso, S, and Campana, F. 2015. "Adjustable velcro® compression devices are more effective than inelastic bandages in reducing venous edema in the initial treatment phase: A randomized controlled trial." *European Journal of Vascular and Endovascular Surgery* 50 (3):368–374. doi:10.1016/j.ejvs.2015.05.014.

Mosti, G, Mattaliano, V, and Partsch, H. 2008. "Influence of different materials in multicomponent bandages on pressure and stiffness of the final bandage." *Dermatologic Surgery* 34 (5):631–639. doi:10.1111/j.1524-4725.2007.34119.x.

Mosti, G, and Partsch, H. 2013. "Bandages or double stockings for the initial therapy of venous oedema? A randomized, controlled pilot study." *European Journal of Vascular and Endovascular Surgery* 46 (1):142–148. doi:10.1016/j.ejvs.2013.04.015.

Mulder, M. 2002. *Basic principles of wound care.* Cape Town: Pearson South Africa.

Nair, B. 2014. "Compression therapy for venous leg ulcers." *Indian Dermatology Online Journal* 5 (3):378–382. doi:10.4103/2229-5178.137822.

Nelson, A. 1995. "Compression bandaging for venous leg ulcers." *Journal of Tissue Viability* 5 (2):57–62. doi:10.1016/S0965-206X(14)80179-4.

Neumann, H A M. 2013. "Elasticity, hysteresis and stiffness: The magic triangle." *Veins and Lymphatics* 2 (1):17–18. doi:10.4081/vl.2013.e6.

Neumann, H A, Partsch, H, Mosti, G, and Flour, M. 2016. "Classification of compression stockings: Report of the meeting of the international compression club, Copenhagen." *International Angiology* 35 (2):122–128.

Orbach, E J. 1979. "Compression therapy of vein and lymph vessel diseases of the lower extremities: A present day overview." *Angiology* 30 (2):95–103. doi:10.1177/000331977903000203.

Partsch, H. 2005. "The static stiffness index: A simple method to assess the elastic property of compression material in vivo." *Dermatologic Surgery* 31 (6):625–630. doi:10.1111/j.1524-4725.2005.31604.

Partsch, H. 2008. "Intermittent pneumatic compression in immobile patients." *International Wound Journal* 5 (3):389–397. doi:10.1111/j.1742-481X.2008.00477.x.

Partsch, H, and Mortimer, P. 2015. "Compression for leg wounds." *British Journal of Dermatology* 173 (2):359–369. doi:10.1111/bjd.13851.

Partsch, H. 2017. "Use of compression therapy." In *Sclerotherapy – Treatment of varicose and telangiectatic leg veins – vi edition*, edited by Goldman, M P, 137–172. China: Elsevier.

Partsch, H, Rabe, E, and Stemmer, R. 1999. *Compression therapy of the extremities.* Paris: Editions Phlebologiques Francaises.

Rajendran, S, Anand, S C, and Rigby, A J. 2016. "Textiles for healthcare and medical applications." In *Handbook of technical textiles: Technical textile applications*, edited by Horrocks, A R and Anand, S C, 135–168. Cambridge, London: Woodhead Publishing. doi:10.1016/B978-1-78242-465-9.00005-7.

Ramelet, A A. 2002. "Compression therapy." *Dermatologic Surgery* 28 (1):6–10. doi:10.1046/j.1524-4725.2002.01181.x.

Rithalia, S V S, Heath, G H, and Gonsalkorale, M. 2002. "Evaluation of intermittent pneumatic compression systems." *Journal of Tissue Viability* 12 (2):52–57. doi:10.1016/S0965-206X(02)80014-6.

Sayegh, A. 1987. "Intermittent pneumatic compression: Past, present and future." *Clinical Rehabilitation* 1 (1):59–64. doi:10.1177/026921558700100112.

Scholl Manufacturing Co Inc. 1964. "Tubular bandage and material therefor." Patent no. US 3306288A. Filed 9/11/1964, and Issued 28/2/1967.

Scott, R W. 1914. "Seamless stocking." Patent no. US 1123402A. Filed 27/5/1914, and Issued 5/1/1915.

Shivers, J C. 1958. "Segmented copolyetherester elastomers." Patent no. US 3023192. Du Pont. Filed 29/5/1958, and Issued 27 Feb 1962.

Sikka, M P, Ghosh, S, and Mukhopadhyay, A. 2014. "The structural configuration and stretch property relationship of high stretch bandage fabric." *Fibers and Polymers* 15 (8):1779–1785. doi:10.1007/s12221-014-1779-2.

Thomas, S. 1990. "Bandages and bandaging: The science behind the art." *Care Science and Practice* 8 (5):56–60.

Thomas, S. 1997. "Compression bandaging in the treatment of venous leg ulcers." *World Wide Wounds*. First publication. doi:10.12968/jowc.1996.5.9.415.

Vowden, K R, Mason, A, Wilkinson, D, and Vowden, P. 2000. "Comparison of the healing rates and complications of three four-layer bandage regimens." *Journal of Wound Care* 9 (6):269–272. doi:10.12968/jowc.2000.9.6.25992.

Vowden, K, Vowden, P, Partsch, H, and Treadwell, T. 2011. "3M™ Coban™ 2 compression made easy." *Wounds International* 2 (1):1–6.

Waldie, J, Tanaka, K, Tourbier, D, Webb, P, Jarvis, C, and Hargens, A. 2002. "Compression under a mechanical counter pressure space suit glove." *Journal of Gravitational Physiology* 9 (2):93–97.

Wallace, H C, and Wilmington, D. 1938. "Linear polyamides and their productions." Patent no. US2130523A. DuPont. Filed 2/1/1935, and Issued 20/9/1938.

William, B. 1848. "Improvements in manufacturing elastic stockings and other elastic bandages and fabrics." Patent no. 12294. Cambridge heath. Filed 26/4/1848, and Issued 26/10/1848.

William, B, and Thomas, H. 1846. "Improvements in the manufacture of articles where india rubber or gutta-percha is used." Patent no. 11455. Filed 10/11/1846, and Issued 19/5/1847.

Williams, C. 2002. "Actico: A short-stretch bandage in venous leg ulcer management." *British Journal of Nursing* 11 (6):398–401. doi:10.12968/bjon.2002.11.6.10132.

Youn, Y J, and Lee, J. 2019. "Chronic venous insufficiency and varicose veins of the lower extremities." *Korean Journal of Internal Medicine* 34 (2):269–283.

Zamporri, J, and Aguinaldo, A. 2018. "The effects of a compression garment on lower body kinematics and kinetics during a drop vertical jump in female collegiate athletes." *Orthopaedic Journal of Sports Medicine* 6 (8):1–6.

2 Materials and Structure of Compression Bandages and Stocking Devices

Rabisankar Chattopadhyay

CONTENTS

2.1 INTRODUCTION

Compression bandages and stockings are extensively used in compression therapy, scar management, venous and bone and muscle injury, lymphatic disorders injury prevention, and performance enhancement. The stockings are elastic knitted fabric tubes, and pressure bandages are stretchable narrow fabrics. Pressure stockings can be directly worn whereas pressure bandages are to be wrapped under uniform tension around the affected limb. Compression is due to development of inward pressure as a result of tension in the bandage or stocking fabric. Pressure stockings are usually custom-made. The first scientific record of the use of pressure stockings was found in the 1970s in the treatment of burn scars and postoperative conditions (Macintyre and Braid 2006, Engrav et al. 2010). Pressure reduces formation of collagen within the scar and also alleviates itchiness and pain. Bandages are used to manage sprains

DOI: 10.1201/9781003298526-2

and diseases such as venous ulcers or lymphedema. Bandage has the advantage that it can easily be wrapped, unwrapped, and rewrapped with different wrapping geometries to accommodate any change in the limb size and shape. In some treatment protocol both bandage and stocking are used. Once the shape and size of the affected limb stabilize using bandage, the compression treatment is performed with stockings. Thus, compression bandages and stockings are complementary (Fanette et al. 2015). There are a plethora of publications on compression stockings, and bandages in research journals dealing with compression therapy and compressive textiles. Material, structural characteristics, properties, pressure development, and treatment efficacy have been the theme of these publications. In this chapter, the focus of the discussion is limited to the material and structural aspects of compression bandage and stockings.

2.2 FUNCTIONAL REQUIREMENTS IN COMPRESSION BANDAGES AND STOCKINGS

The material and structure are decided keeping in mind the functional requirements of the bandage and stockings. There are plenty of materials in the form of fiber (natural and synthetic), yarns (spun, filament, blended, textured, composite), and fabrics (woven, knitted, nonwoven, braided, spacer, etc.) available for developing compressive textile. It is necessary to know the requirements of such products so that the selection of fiber or fiber mixtures, fabric or fabric combinations, finishes, and fabric joining techniques can be objectively determined. The environment in which the product is to be used, patient's physiology, age, and duration of use also play an important role. The functional needs in compressive bandages/stockings are:

- Development of designated pressure
- Maintenance of pressure over a prolonged time
- Comfortable to wear
- Resistant to slippage
- Quick doffing and donning
- Anti-bacterial
- Permeable to moisture to avoid rashes, itching, and odor

The needs enumerated above may vary from person to person depending upon the type of disease, age, sex, and body physiological parameters. Each need is a complex function of fiber, yarn, fabric structure, finishing, and joining techniques. A variety of combinations of these parameters are possible to offer an optimum solution to meet the performance requirements. As a result, multiple types of compressive bandages and stockings are commercially available.

2.3 MATERIAL FOR COMPRESSION BANDAGES AND STOCKINGS

Since stretchability is the fundamental property required in compression bandages and stockings, the fiber or yarn selection will be dependent on the amount of stretch required in fabric, keeping in mind ease of donning and pressure required. Pressure being a function of tension in the constituent yarns, the stress strain property of the fiber or yarn becomes important.

Stretchability being a fundamental requirement in the product, it is important to know the factors responsible for stretch generation. The three important factors are (i) stretchability of fibers, (ii) type of yarn (textured, core spun "Lycra™", rubber thread), and (iii) fabric with the right constructional geometry. These factors can be manipulated to produce fabrics with a wide range of stretchability.

2.3.1 Fiber and Yarn

The obvious choice of fibers are nylon, polyester, elastomeric fiber (spandex/"Lycra™"), and natural rubber for imparting stretchability and cotton, viscose rayon, and acrylic for comfort. Typical properties of these fibers are stated in Table 2.1. Nylon filament is mainly used as it has excellent tenacity (4.6–5.8 g/denier), elastic recovery (100% recovery up to 8% extension), and 4–4.5% moisture regain. The ability to absorb some moisture makes it comfortable in comparison to polyester. The nylon percentage can vary between 50% and 85% in the final product.

Cotton or viscose rayon/polynosic fibers may be used in addition to nylon to improve the comfort properties further, as these fibers' moisture regain are much more than nylon. Elastomeric fibers such as rubber or spandex, with a fiber content of 15–34%, are incorporated to increase the elasticity. Natural rubber is not used much since it is difficult to color and is not available in the desired fineness.

The fiber composition, construction details, and physical parameters of pressure stockings are stated in Table 2.2 based on the information available in the literature (Macintyre, Baird and Weedall 2004, Ghosh et al. 2008, Hui and Ng 2001, Hui and Ng 2003, Gupta, Chattopadhyay and Bera 2011).

Fine (20 denier) spandex filament is used along with nylon for weft-knitted structures. It improves the length-wise extensibility (Chattopadhyay, Gupta and Bera 2011). Cotton yarn is used along with nylon for comfort and feel. Covered spandex (320–595 denier) or rubber yarns (more than 1500 denier) are inlaid into the structure for stretch. Since spandex and rubber are not skin-friendly, they are first covered by a multifilament nylon yarn (40–50 denier) to improve tactile comfort especially when heavy denier spandex or rubber is used. Usually, two layers of nylon are wrapped onto the elastomeric yarn in opposite directions (Figure 2.1). The two wrappings give a balanced structure with a minimum of twist liveliness.

A survey on spandex threads used in different fitting garments reveals that 70–140 denier spandex is normally used in fitted garments, while heavy spandex yarn (> 420 denier) is used for pressure stockings and foundation garments (Ibrahim 1968).

The linear density of spandex/rubber can be chosen according to the pressure requirement. The 50–360 denier nylon filament yarn is used. Textured nylon filament is preferred to multifilament nylon as it improves bulk and extensibility (Hui and Ng 2003, Hui and Ng 2001, Gupta, Chattopadhyay and Bera 2011).

TABLE 2.1
Properties of Fiber used in Compressive textile

Properties	Cotton	Wool	Viscose	Nylon	Polyester	Acrylic	Spandex
Deniersity (g/cm^3)	1.54	1.32	1.52	1.14	1.36	1.14	1.10-1.35
Tenacity (cN/tex)	20–50	10–16	18–35	40–60	25–60	20–35	4–12
Breaking extension (%)	6–10	25–50	15–30	20–60	15–50	80–95	400–800
Elastic recovery at 10% extension	Nil	45–50	–	90–95	50–80	55–80	100
Modulus (cN/tex)	300–600	150–300	200–300	50–300	25–400	300–500	0.3–0.7
Moisture absorption (%)	7–11	15–17	11–14	3.5–4.5	0.2–0.5	1.0–2.0	0.15–1.5

TABLE 2.2
Constructional Details of Pressure Stocking Reported in the Literature

Knit construction	Constituent fiber: Loop formation	Constituent fiber: Inlaid	Fabric thickness (mm)	Fabric areal density (g/m²)
Single jersey	Nylon (85%)	Spandex (15%)	0.77	250
Rib	Nylon (65%), Cotton (15%)	Rubber (20%)	2.61	714
Rib	Nylon (76%)	Rubber (24%)	1.42	768
Weft knit	Cotton (82%)	"LycraTM" (18%)	1.16	570
Warp knit	Nylon	Spandex	0.55	220
Warp knit	Nylon (50%), Cotton (16%)	Spandex (34%)	0.7	270
Power net	Nylon (60–88%)	Spandex (40–12%)	0.39–0.51	195–276
Warp knit	Nylon (48–60%) Cotton (19–28%)	Spandex (11–31)	0.53–0.61	175–145
Rib knit	Nylon (90%)	Spandex (10%)	1.33	424
Single jersey	Nylon (72.2%)	Spandex (27.8%)	1.02	396

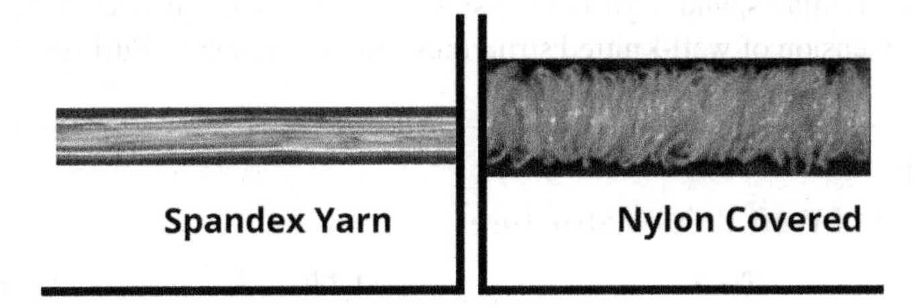

FIGURE 2.1 Bare and nylon-covered spandex yarn.

2.3.2 Fabric

The constructional details of pressure stockings reported by different authors in the literature have been compiled in Table 2.2. The fabrics are usually warp- or weft-knitted. In the case of weft-knitted structure, rib, single jersey, and pique structures are used (Macintyre, Baird and Weedall 2004, Ghosh et al. 2008). Nylon along with fine spandex yarn form loops and spandex or rubber threads are inlaid to make the fabric stretchable.

A typical image of the single jersey structure is shown in Figure 2.2. Single jersey structures are light and thin, whereas rib knits are thick and heavy fabrics. This

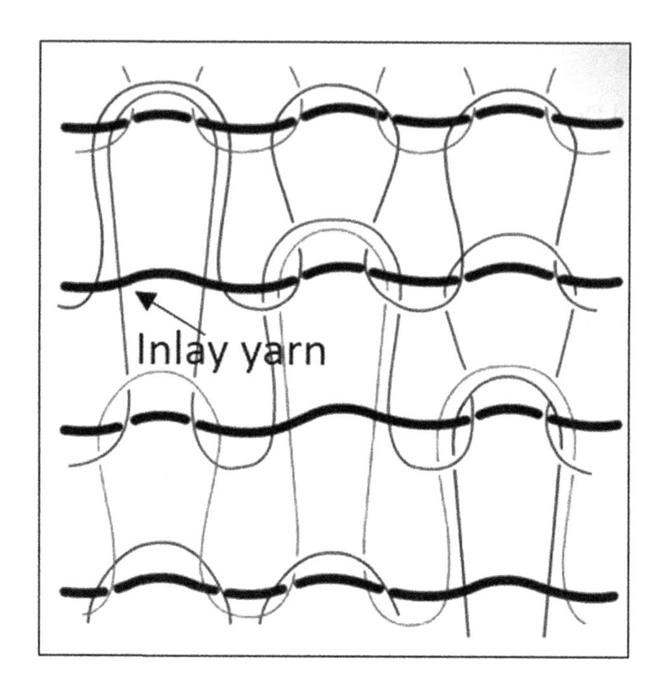

FIGURE 2.2 Single jersey structure.

makes rib structures warmer and uncomfortable in hot weather. But they are used when better insulation is required. The Pique structure has a combination of loop and tucks knit. Tuck knit improves the widthwise extension of the structure, so pique structures with inlay spandex yarn are also sometimes used (Spencer 2001). A comparison of extension of weft-knitted structures (Table 2.3) shows Purl structure leads

TABLE 2.3
Comparison of Weft-Knitted Structure

Property	Plain	1×1 Rib	1×1 Purl
Appearance	Different on face and back V-shape on the face and arc on the back	Same on both sides, like the face of plain	Same on both sides like back of plain
Extension			
Lengthwise	10–20%	10–20%	50–100%
Widthwise	30–50%	50–100%	30–50%
End uses	Ladies' stockings, fine cardigan, men's and ladies' shirt, dresses, base fabric for coating	Top of socks, cuffs, waist band, collars, men's outwear, knitwear, underwear	Children's clothing, thick and heavy outwear

Note: Moderate extension: 10–20%, high extension: 30–50%, and very high: 50–100%

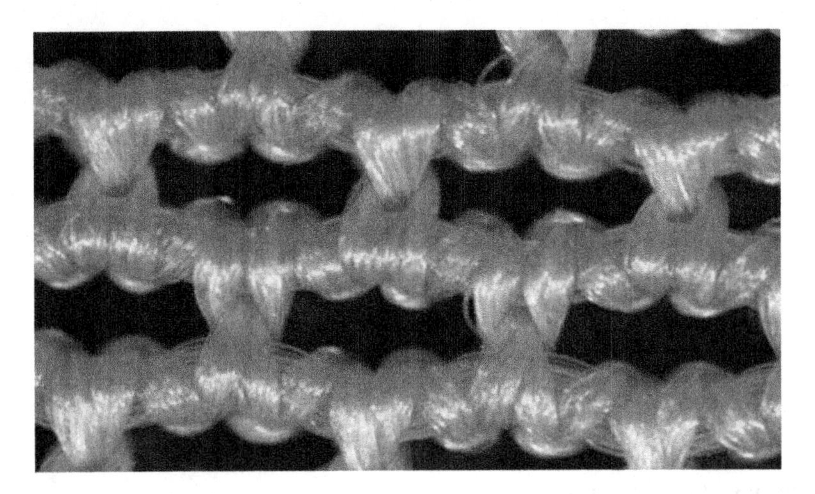

FIGURE 2.3 Power-net fabric.

to maximum extensibility both in length and weft-wise directions. The loops are made of nylon filaments and the elastic component is inlaid. Depending upon the stretch, recovery, and tension required, the proportion of the two components varies.

In warp-knitted structures, the power net structure is very popular (Figure 2.3). It remains springy when stretched due to the spandex component. Power net fabrics are stretchable by 50% in both longitudinal and transverse directions.

Fabric thickness usually varies from fine (0.5 mm) to thick (2.6 mm). Similarly, fabric areal density also varies within a wide range of 220–768 g/m^2. Low thickness and low fabric weight lead to better comfort, but pressure generation may be poor. Thicker heavy weight fabric on the contrary will offer more pressure but a feeling of discomfort, especially in summer. So, one has to choose an optimum combination according to the requirement.

2.4 BANDAGES CONSTRUCTION

Kumar et al. (2014) and Kankariya et al. (2021) have made excellent reviews on compression bandages. The construction of the bandage depends upon the type of bandage to be produced. Stiffness (resistance to deformation) is an important criterion in bandage selection. Inelastic short-stretch, multiple component (short stretch), and multiple component (multilayer) bandages fall into the category of highly stiff, whereas long-stretch bandages fall into the category of low stiff.

In a multilayer structure, the same bandage is applied in an overlapping manner on the affected area of the limb. In a multiple-component structure, multiple types of materials are combined into a single compression bandage. The popular multiple-component four-layer bandage (4LB) consists of a padding bandage (wadding or orthopedic wool), a crepe bandage, an elastic bandage, and an outer layer made of an elastic cohesive bandage (Christopher, Kellie and Ruth 2015). The properties of each layer differ from each other.

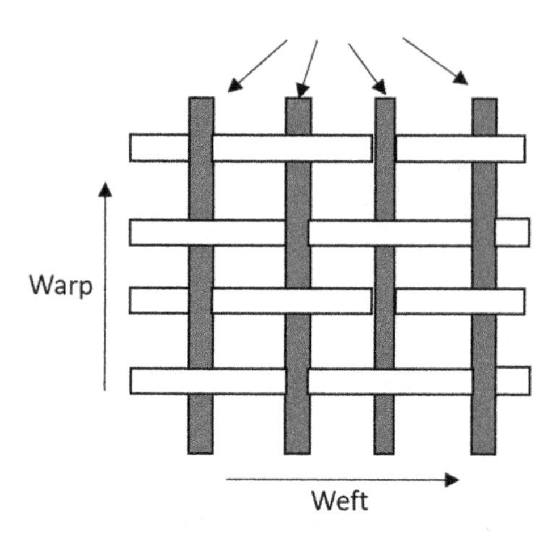

FIGURE 2.4 Woven bandage.

The padding bandage is placed next to the skin and below the compressive bandage. The purpose of the padding bandage is to evenly distribute the pressure by absorbing high pressure at the tibia and fibula regions. The padding bandage is nonwoven material.

P-LA-C-E ("Pressure level – Layers – Components of material – Elasticity) classification of multicomponent bandage system available commercially has been reported (Kankariya et al. 2021). Information on components, structure, material used, and elastic nature of the commercial bandages has been stated.

A bandage could be of knitted or woven construction. The knitted construction gives inherent stretchability. Highly twisted yarns are used to enhance the stretch property. Woven construction consists of two sets of orthogonal yarns known as warp and weft yarns that are interlaced (Figure 2.4). In knitted construction, the yarns are looped around each other. As a result, the structure can easily deform and conform to the complex body shape. The stretchability of the fabric and its covering power depends upon the courses and wales per unit length. Knitted fabrics are categorized as warp-knitted and weft-knitted. In weft-knitted fabric, the yarn follows the path across the length of the fabric. In warp-knitted fabric, the yarn path moves along the length of the fabric.

2.5 STOCKING CONSTRUCTION

Stockings are classified according to the pressure they generate, i.e., low (14–17 mmHg), medium (18–24 mmHg), and high (25–35 mmHg). The material selection and construction must result in the development of right pressure.

Compression stockings are made in two varieties (i) elastic compression stockings and (ii) ulcer kits (i.e., double stockings). Stocking is a seamless, ready-to-wear one-piece garment or may be stitched according to the applied reduction factor and

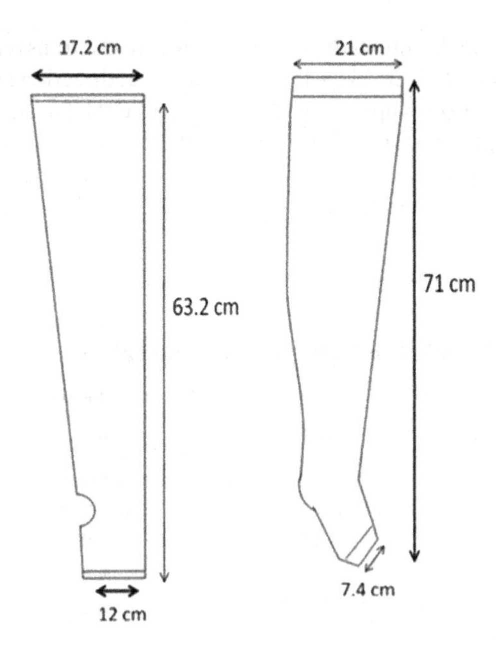

FIGURE 2.5 Compression stocking.

patients' body dimension. Some stockings are also designed to be adjustable, to meet the requirement of changing body dimensions as the therapy progresses. The design is crucial for successful pressure therapy. If the stocking does not follow the body contour, then it cannot generate the right pressure (Bera, Gupta, and Chattopadhyay 2010). Patterns have to be developed keeping the complex geometry of the body part in mind. A typical outline of compression stockings is shown in Figure 2.5.

As can be seen in Figure 2.5, stockings are tubular-knitted seamless garments of thigh length. They may have a tapered configuration or a more contoured shape corresponding to the lower part of the leg. A slit at the ankle position allows the heel to pass through. The top and bottom ends are folded over and hemmed. A special elastic band may be attached to the top and coated with silicone rubber for a better grip.

Velcro fasteners are used for self-adjusting pressure by controlled tightening. They are the only inelastic compression devices. The disadvantages are bulkiness and impose restrictions on mobility.

2.6 PROPERTIES OF BANDAGE AND STOCKING FABRICS

2.6.1 Tensile Property

The tensile property is important as it is directly related to pressure generation. Anand (2010) has reported that the breaking load of stocking varies from 16 to 30

Kgf in the longitudinal direction and 6.6 to 12.2 Kgf in the transverse direction. The breaking elongation is 155% to 332% in the longitudinal direction and 220% to 400% in the transverse direction. Gupta et al. (2011) compared the physical and mechanical properties of three pressure stockings (Table 2.4). The breaking load and extension of the stockings lie in the range of 15–23 Kgf and 755–1,350%, respectively. The force-extension curves of the three stocking fabrics are shown in Figure 2.6. The

TABLE 2.4
Construction and Tensile Properties of Stocking Fabrics

Sl. no.	Fabric parameters	Pressure stockings		
		PG_1	PG_2	PG_3
1	Fiber content	Nylon/Rubber/Cotton (65:20:15)	Nylon/Rubber/Spandex (76:18:6)	Nylon/Spandex (85:15)
2	Fabric weight (g/m^2)	714	768	250
3	Thickness (mm)	2.61	1.42	0.77
4	Knit construction	1×1 rib	1×1 rib	Single jersey
6	Elastic yarns/inch	14	24	27
7	Wales/inch	36	44	30
8	Courses/inch	14	24	27
9	Loop length (mm)	0.6	0.4	0.2
10	Tightness Factor	9	14	32
11	Tensile strength (Kgf)	20	23	15
12	Breaking extension (%)	1,350	1,215	755

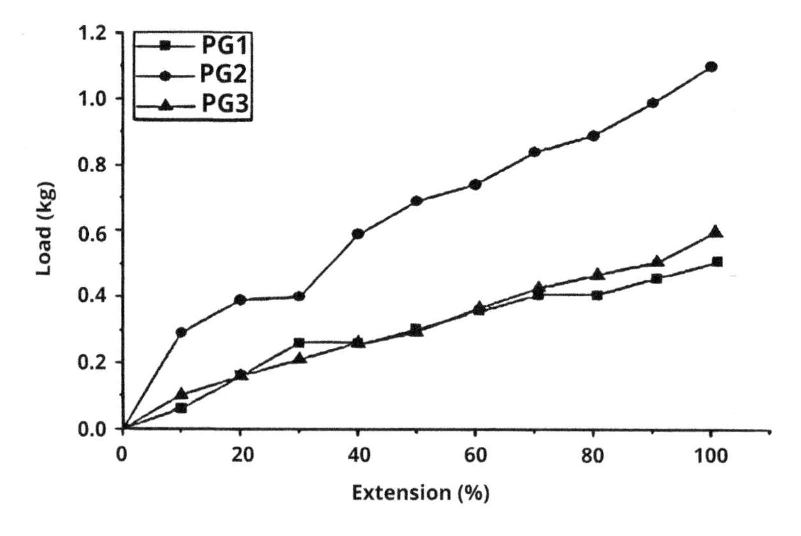

FIGURE 2.6 Force–extension diagram of pressure stockings.

nature of force-elongation curves is mainly governed by the intrinsic properties of constituent elastic filaments and their number per unit length of the fabric. The force-extension characteristics of PG1 and PG3 overlap each other, even though PG3 is lighter in weight than PG1. PG2 shows higher force at any extension level. It appears that force-elongation characteristics of the Spandex yarn used in PG3 are superior to that used in PG1.

Sikka et al. (2014) reported that the bandages with coarser elastic yarns exhibit low fabric growth, high breaking load, and high elastic recovery. The sample with the maximum number of elastic yarns and a tighter structure develops a much higher tension force at a low stretch. The ability to apply high pressure with minimum stretch is achieved in the case of knitted bandage fabrics, with a lesser number of high-stretch yarns as compared to woven samples.

Manawacharitha et al. (2015) observed that the elastic modulus of the bandages before lock-out increases with an increasing elastomeric yarn end density. In bandages with two types of warp, non-stretch yarns do not significantly affect the load/elongation behavior before lock-out if they have a high percentage of crimp. Beyond the lock-out region, the amount of non-stretch yarns as well as the warp and weft end densities affects the elastic modulus. For bandages consisting of stretch and non-stretch yarns, the crimp of non-stretch yarns affects total elongation, elastic modulus before lock-out, and the width of the lock-out region.

Muhammad et al. (2016) observed that by increasing the elastane count, both fabric stretch and recovery percentage along the warp increased, whereas in the case of weft direction only recovery percentage increased. An increase in thread density decreased fabric contraction, and stretch percentage along warp and weft. An increase in the float length increased fabric contraction, stretch along both the warp and weft and fabric warp and weft-way recovery. Fabric with a heavier elastane count, more thread density, and shorter float length leads to higher pressure. Knitted fabrics demonstrated much higher stretch and recovery percentage than woven bi-stretch fabrics. Woven bi-directional stretch fabrics exhibit better compression properties before and after washes and retain their durability after repeated use, whereas knitted stretchable fabrics lose their compression ability after repeated use (Muhammad et al. 2017).

2.6.2 Frictional Property

The frictional properties of the stocking are very important as it is used next to the skin in scar management. The frictional properties of human skin depend not only on the nature of the skin (texture, suppleness, smoothness, dryness, or oiliness) but also on its interaction with clothing. Zhanj and Cheung (2006) mentioned the frictional coefficient of skin at different positions of the body and also interaction with different fibers. The frictional coefficient of the skin on hand, forearm, and leg is almost the same ranging between 0.40 and 0.47 with an average of 0.41 ± 0.14. Nylon shows the lowest coefficient of friction against human skin. Weft-knitted structure has the highest friction, followed by power net structure (Macintyre, Baird and Weedall 2004).

2.7 INFLUENCE OF MATERIAL AND STRUCTURE ON PRESSURE DEVELOPMENT AND DECAY

The four main influencing factors that affect pressure development are (i) body characteristics, (ii) intrinsic properties of the material (fiber and fabric), (iii) fabric constructional parameters, and (iv) fit. The body characteristics include the body curvature and compliance of the body part on which the stocking is worn. The material properties are related to the tensile property (load elongation characteristics) of fibers and yarns and their fineness. The fabric constructional parameters are loop density, ends and picks per unit length, and the number of layers and their constituents. The fit is the reduction factor in stockings.

The tension at a given strain in the fabric depends upon its stress–strain relationship, elastic modulus, and stretch applied during wrapping. The intrinsic property of the constituents plays a dominant role. Stretchable fiber will also make the product also stretchable, whereas non- stretchable fiber will make it stiff. Thus, at a given stretch more pressure will develop when non-stretchable fibers are used.

Weft-knitted structure extends more than warp-knitted structure, and as a result, it develops less tension and thereby less pressure. The opposite is true for the warp-knitted structures. In the case of woven structure, made from highly twisted yarns, the level of twist influences the stretchability and pressure development.

In multilayer bandages, the final pressure is the summation of transverse forces generated by each layer. Thus, the number of wrapping layers will have a significant effect on pressure. Increasing the number of layers leads to a more coherent and supportive structure, and therefore generates high pressure. Applying a bandage with a 50% overlap effectively produces two layers of bandage and generates a pressure double than that produced by a single-layer bandaging system. If the bandage is wrapped under constant tension or force, the pressure developed will be less as the same force will be applied over a large surface area. Das et al. (2010) studied the effect of bandage width and number of layers using a leg-segment prototype and obtained similar results.

The fit is related to the reduction factor. By definition, the reduction factor (R_F) is:

$$R_F = \frac{\text{Body circumference}(\text{m})\text{-stocking circumference}(\text{m})}{\text{Body circumference}(\text{m})} \quad (2.1)$$

Stockings are always smaller than the body parts on which they are worn. As a result, they remain in an extended state. The extension depends upon the reduction factor. The fabric strain is given by Equation 2.2.

$$F_s = \frac{\text{Body circumference-stocking circumference}(\text{m})}{\text{stocking circumference}(\text{m})} \quad (2.2)$$

The theoretical relationship between R_F and F_s is:

$$F_s = \frac{R_F}{1 - R_F} \tag{2.3}$$

The reduction factor helps in developing customized stockings suggested by the occupational therapists according to the needs of the patients. Pressure garments are normally fabricated based on a standard reduction factor of 10, 15, or 20%. The reduction factor used is constant and is not normally changed based on the dimensions of the body part or the specific properties of the fabric. Surveys conducted (Brown, Burns 2001, Macintyre and Baird 2005) have shown that 10–20% R_F is generally maintained during the manufacturing of stockings for burn patients.

Since pressure is a function of tension, at a given reduction factor, the tension level changes with a change in the intrinsic properties of the elastic inlay yarn, its linear density (denier), and numbers per unit width in the stocking. Therefore, high loop density and heavy denier elastic thread and a high initial modulus will lead to higher pressure.

The components of compression bandages being made from polymeric materials are viscoelastic in nature. Due to stress relaxation phenomenon, the tension in the constituent polymeric threads declines over time. Because pressure is a function of fabric tension, it is expected to decay with time. According to Hui and Ng (2003), the pressure at any time "t" is:

$$P = P_a + \left(P_0 - P_a\right)e^{-\alpha_P t_P} \tag{2.4}$$

where P = pressure (N/cm^2), P_a = pressure (N/cm^2) in the fabric as "t" tends to infinity, P_0 = Initial pressure (N/cm^2), α_P = pressure decay (N/cm^2/min). In a 24-hour duration study, the pressure decay has been observed to continue for the first 30 minutes before stabilization.

Giovanni and Hugo (2010) reported a loss of pressure over time in the case of inelastic compression bandage. The loss starts immediately after application and tends to stabilize after 24 hours. In contrast, the elastic compression bandage tends to maintain its pressure over time and shows only a very small and insignificant pressure loss after 1 week.

Stockings need to be frequently washed as it is worn for a prolonged period. The pressure exerted by the garments decreases with every washing cycle. This behavior is not acceptable in the treatment of chronic diseases. Ideally, there should be no deformation (Ghosh et al. 2008).

2.8 CHALLENGES

Bandages are typically used for healing and stockings for preventing recurrence. There are many challenges to overcome. The compression bandage should be designed in such a way that the differences in the bandager's wrapping skill do not seriously affect the target pressure range. The force-elongation characteristics of the component materials chosen become very important here. The force developed due to stretch during wrapping should be insensitive to the elongation variation. The bandage has to be engineered through the proper selection of raw materials, their assembly, and finishing techniques. Consistency in sub-bandage pressure over a long-term application is another area to be looked into. Polymeric materials used in the manufacturing of the bandages are prone to creep deformation, which causes the bandage to become loose with gradual reduction in pressure over time.

Sizing of stockings is another area of concern. The stocking must fit the patient properly. A given size bought off the shelf may not fit everybody well, as actual size of the affected limb varies a lot between people. Misfit stockings will prolong healing.

The compression stockings must balance between functionality and comfort. Discomfort may arise due to low or too high thermal insulation and improper moisture permeability. The climatic condition varies throughout the year and also between countries. Thus, country-specific design may be needed.

2.9 CONCLUSION

A brief review on the material, components, and structure of pressure bandages and stockings has been presented. Pressure bandages and stockings are highly engineered products used by medical professionals for managing burn scars, venous ulcers, muscular strains, and other diseases. It has been shown that not only material but also the structure and constructional parameters of both fabric and product architecture play an important role. The fibers and structure are also key parameters for maintaining comfort and pressure over a prolonged time. The main challenge lies in the design and selection of material that ensures consistency in pressure over time and overall physiological comfort in a variety of climatic conditions.

REFERENCES

Anand, S C. 2010. "A study of the modelling and characterization of compression garments for hypertrophic scarring after burns." Proceedings, 39th Textile Research Symposium, IIT Delhi, New Delhi, India.

Bera, M, Gupta, D, and Chattopadhyay, R. 2010. "Studies on pressure stockings." In International conference of healthcare and hygiene textiles and clothing. HEAT, Coimbatore, India 11.

Brown, C A. 2001. "A comparison of the outcomes of two clinical audits of burn pressure garment satisfaction and compliance in Saudi Arabia." *Burns* 27 (2001):342–334.

Chattopadhyay, R, Gupta, D, and Bera, M. 2011. "Effect of input tension of inlay yarn on the characteristics of knitted circular stretch fabrics and pressure generation." *Journal of the Textile Institute* 103 (6):636–642.

Christopher, L A, Brown, K R, and Ruth, L B. 2015. "Compression therapies for chronic venous leg ulcers: Interventions and adherence." *Chronic Wound Care Manag Res* 2 (2015):11–21.

Das, A, Alagirusamy, R, Goel, D, and Garg, P. 2010. "Internal pressure profiling of medical bandages." *Journal of the Textile Institute* 101 (2010):481–487.

Engrav, L H, Heimbach, D M, Rivara, F P, Moore, M L, Wang, J, Carrougher, G J, Costa, B, Numhom, S, Calderon, J, and Gibran, N S. 2010. "12-Year within-wound study of the effectiveness of custom pressure garment therapy." *Burns* 36 (2010):975–983.

Fanette, C, Martin, F R, Badel, P, Reynald, C, Pascal, G, and Molimard, J R. 2015. "Experimental investigation of pressure applied on the lower leg by elastic compression bandage." *Annals of Biomedical Engineering* 43 (12):2967–2977.

Ghosh, S, Mukhopadhyay, A, Sikka, M, and Nagla, K S. 2008. "Pressure mapping and performance of the compression bandage/garment for venous leg ulcer treatment." *Journal of Tissue Viability* 17 (2008):82–94.

Giovanni, M, and Partsch, H. 2010. "Inelastic bandages maintain their hemodynamic effectiveness over time despite significant pressure loss." *Journal of Vascular Surgery* 52 (4):925–931.

Gupta, D, Chattopadhyay, R, and Bera, M. 2011. "Compression stockings-structure property analysis." *Asian Textile Journal* 29 (1):39–45.

Hui, C L, and Ng, S F. 2001. "Model to predict interfacial pressures in multilayer elastic fabric tubes." *Textile Research Journal* 71 (8):683.

Hui, C L, and Ng, S F. 2003. "Theoretical analysis of tension and pressure decay of a tubular elastic fabric." *Textile Research Journal* 73 (3):268.

Ibrahim, S M. 1968. "Mechanics of form-persuasive garments based on spandex fibers." *Textile Research Journal* 38:950.

Kankariya, N, Laing, M R, and Wilson, C A. 2021. "Textile-based compression therapy in managing chronic edema: Complex interactions." *Phlebology* 36 (2):100–113.

Kumar, B, Das, A, and Alagirusamy, R (Eds). 2014. In Science of compression bandage. New Delhi, India: Wood Head Publishing India Ltd.

Macintyre, L, and Baird, M. 2006. "Pressure garments for use in the treatment of hypertrophic scars: A review of the problems associated with their use." *Burns* 32:5–10.

Macintyre, L, and Baird, M. 2005. "Pressure garments for use in the treatment of hypertrophic scars – An evaluation of current construction techniques in NHS hospitals." *Burns* 31 (2005):11–14.

Macintyre, L, Baird, M, and Weedall, P. 2004. "The study of pressure delivery for hypertrophic scar treatment." *International Journal of Clothing Science and Technology* 16 (2):173–183.

Manawacharitha, R A, Wimalaweera, W A, and Wijeyaratna, S M. 2015. "A study about the lock-out point of bandages used for compression therapy." *Journal of Engineering and Technology* 3 (2):1–12.

Muhammad, M, Tanveer, H, Malik, M H, and Yasir, N. 2016. "Modeling the effect of elastane linear density, fabric thread density, and weave float on the stretch, recovery, and compression properties of bi-stretch woven fabrics for compression garments." *The Journal of The Textile Institute* 107 (3):307–315.

Muhammad, M, Yasir, N, Jawairia, U, Muhammad, U, and Khubab, S. 2017. "Comparison of compression properties of stretchable knitted fabrics and bi-stretch woven fabrics for compression garments." *The Journal of The Textile Institute* 108 (4):522–527.

Sikka, M P, Ghosh, S, and Mukhopadhyay, A. 2014. "The structural configuration and stretch property relationship of high stretch bandage fabric." *Fibers and Polymer* 15 (2014):1779–1785

Spencer, D J. 2001. *Knitting technology* (3rd edition). Cambridge: Woodhead Publishing Limited.

Zhanj, M and Cheung, JT 2006. "Mechanics of human skin and underlying soft tissue." In *Biochemical engineering of textile and clothing*, edited by Li, Y, and Dai, X-Q. Cambridge, England: The Textile Institute, pp. 111–123, CRC Press,Woodhead Publishing Ltd.

3 Mechanical Properties of Compression Garments

Wan Syazehan Ruznan, Mohd Khairi Mohd Yusoh, Fazieyana Mohd Tohid, and Nur Ain Yusof

CONTENTS

3.1 INTRODUCTION

The use of compression is important in the care of wounds and injuries such as edema and sprains to promote healing and to prevent further harm to the wounds and injuries. The effects of compression therapy can be influenced by the pressure on the limb and the stiffness of the bandage characterizing the elastic property of the bandage fabrics (Mosti and Mattaliano 2007).

3.2 EFFECT OF ELASTICITY ON COMPRESSION

Stress relaxation is a type of transient situation that happens in real life on viscoelastic materials. Generally, experiments were conducted to observe how the material deformed and to calculate the amount of stress required to maintain that deformation at a constant value. Meanwhile, relaxation time is commonly used to describe the stress relaxation behavior of a viscoelastic material, which indicates the time required to progress to a new relaxed state of the material (Geršak, Šajn and Bukošek 2005, Thomas 2003).

The efficacy of compression bandages depends on the mechanical properties of the bandage fabric itself, the user's movement, the pressure applied on the fabric surface, and the surrounding situation (Ruznan et al. 2021). Some previous studies explore the elasticity properties of the bandage fabric by simulating it on mathematical models by applying the interface pressure. The result obtained only focuses on

DOI: 10.1201/9781003298526-3

the measurement that results from the reaction of the fabric during stress over time (Ruznan et al. 2021).

Studies on extension properties using different types of parameters have been conducted by a few researchers (Table 3.1). In a study performed by Sikka et al. (2014), the extension level of a woven and knitted sample of bandage fabric was tested by hanging for 30 minutes with a load of 1.8 kg. The properties of bandages such as elastic recovery, maximum stretching, and fabric growth were recorded after 1 hour of relaxation time. The results showed that the density of yarn influenced the value of the maximum extension, as well as higher recovery for both types of samples. Knitted bandages with a smaller number of high-stretch yarns can apply high pressure with minimal stretch as compared to woven bandages (Sikka et al. 2014).

Ardakani, Asayesh and Jeddi (2015) have investigated that the stress relaxation behavior of the fabric could be changed due to the stretch done by the exerted load applied during the application. Three types of warp-knitted fabrics (i.e., locknit, satin, and loop raised) were applied with 5% and 25% extension in wale and course directions, respectively. The samples were extended in both directions for 30 minutes. The stress of the sample fabrics reduces significantly in the first 3 minutes and then gradually diminishes as the stress relaxation test duration approaches constant tension. It was found that the fabric structure has become a substantial factor in fabric stress values and stress relaxation percentages over time, as well as the increase of extension value, which will lead to higher stress (Ardakani et al. 2015).

Fabric structure has an important effect on stress relaxation of fabrics, with most studies investigating stress relaxation of different fabric structures rather than other fabric physical properties (e.g. Hashemi et al. 2016, Ardakani et al 2015, Maqsood et al. 2016, Maqsood et al. 2017).

Besides fabric structure and construction, the viscoelasticity of fabrics with increased elasticity is also determined by the linear density of the elastane, yarn count, and the twist of the yarn (Su, Maa and Yang 2004, Zou 2012). Increasing the count of elastane yarn showed increased stretch and recovery percentage in woven fabrics (Maqsood et al. 2016).

3.3 MULTI-AXIAL STRESS RELAXATION

Garment appearance is adversely affected if the fabric does not fully recover after multi-axial stress relaxation. This situation prompted investigations of multi-axial stress relaxation and the phenomenon of bagging. Both instrumental methods and wear trials have been conducted to measure multi-axial extension. Selected test conditions of various standard and non-standard test methods of multi-axial testing are given in Table 3.1.

A number of ball attachments were developed to be used with Instron testing machines (American Society for Testing and Materials 1980). In the ball burst method, a steel ball attached to a tensile test machine is pushed multi-axially through the specimen that is held rigidly in a circular clamp. Diameters of the ball and test area vary from 6.25 mm and 25.00 mm (British Standards Institution 1968), 25.40

TABLE 3.1
Test Parameters for Measuring Multi-Axial Extension

	Apparatus	Specimen diameter (mm); test area (mm)	Ball diameter (mm)	Pre-tension	Test speed (mm/min)	Limits	Author(s) (year)
Woven: ball burst attachment							
	Instron tensile tester	56; not given	48	0.2, 0.3	20	5 cycles, 0–12 mm extension	Zhang, Li, Yeung and Yao (2000)
Also suitable for knits, nonwoven, felted fabrics	Constant-rate-of-extension (CRE) tensile tester	Not given; 45	25		300	To failure	AS 2001.2.19-1988
	Constant-rate-of-extension (CRE) tensile tester	125; 44.5	25.4		305	To failure	ASTM D 3787-80a
	Celanese bagging tester (attached to Instron)	254; 203.2	101.6		254	Cycled 200–6,800 g load and hold for recovery for 1 minute	Thomas (1971)
Woven: other							
Also suitable for knits, nonwoven, laminated fabrics	Hydraulic bursting tester	Not given; 79.8			100–500 cm^3/minute (increase in volume)	To failure	ISO 13938-1:1999
Also suitable for knits, nonwoven, laminated fabrics	Pneumatic bursting tester	Not given; 79.8			Speed to enable time to burst of 20 seconds	To failure	ISO 13938-2:1999
	Hydraulic bursting tester	75; not given			305	To failure	ASTM D 3786-87
Leather							
	Tensile tester; ball burst attachment	44.5; not given	6.25		12	To failure	BS 3379:1976
	Tensile tester; ball burst attachment	44.5; 25	6.25		12	To failure	BS 3144-8:1968

mm and 44.50 mm (American Society for Testing and Materials 1980), to 101.60 mm and 203.20 mm (Thomas 1971).

The study conducted using a "Celanese" bagging instrument developed in the early 1970s is another type of ball burst attachment for a tensile tester. However, it has larger dimensions (diameter of the ball = 101.60 mm) (Thomas 1971).

3.4 LAYERING EFFECT

The multilayer system is usually used in compression bandaging to give a sufficient amount of pressure to the injury. Multilayer bandaging system with a cohesive long stretch bandage as outer layer demonstrates a smaller pressure decrease (4.0–6.0 mmHg) than the outer systems (10.0–18.0 mmHg) (Hafner, Botonakis and Burg 2000). Four different multilayer bandage systems (i.e. short-stretch, medium-stretch, long-stretch, inelastic) tested on ten healthy participants (five men, five women, all aged between 26 and 65 years) during postural changes, walking and over two days of wear time also found a decrease in interface pressure over time (Hafner, Botonakis and Burg 2000). The interface pressure was measured using an electro-pneumatic device located at 12 points of the leg (Hafner, Botonakis and Burg 2000). The points of measurement were: three points along the medial side of the leg, two points along the lateral side, one point at the medial and lateral retromalleolar space, and one point at the distal dorsum of the foot, over the wrist, at the pretibial area, at the Achilles tendon, and at the calf. Highly elastic (long-stretch) multilayer bandage systems demonstrated the smallest pressure loss over several days of wearing as compared to other bandage systems (Hafner, Botonakis and Burg 2000). An interface pressure of 35–45 mmHg was chosen to examine different multilayer bandaging systems at rest and during exercise. Short-stretch and inelastic bandage systems generated large pressure waves during movement and showed marked pressure loss when patients were in resting position, and more pressure decrease during the first hour of wearing (Hafner, Botonakis and Burg 2000).

Other works also show the same indicator where marked pressure decreases from the standing to the supine position with the use of inelastic bandages (Partsch 1991, Callam, Haiart and M 1991, Veraart, Daamen and Neumann 1996).

3.5 COMPRESSION EFFICACY

Application of a bandage over time may reduce the compression to the lower leg due to stress relaxation behavior of bandage materials (Moffat et al. 2012, Hafner, Botonakis and Burg 2000, Kumar, Hu and Pan 2016). Studies related to bandage fabrics were done mostly on the effect of applied pressure on the body part, without taking into account the relationship to multi-axial stress relaxation of bandage fabrics over time. For instance, studies showed that pressure decreased over time due to a reduction in swelling (Mosti, Picerni and Partsch 2012, Mosti and Partsch 2013). The bandage fabrics may relax after a period of time due to the viscoelastic property of the materials after the reduction of the swollen limb. Pressures lower than 10 mmHg have been shown to prevent evening leg edema

(Partsch, Partsch and Braunc 2006), and stockings with 20 mmHg demonstrated equivalent effectiveness in the reduction of leg swelling after two days in patients with chronic edema, when compared with 60 mmHg bandages (Mosti and Partsch 2013). However, the bandage pressure on the lower leg decreased over time, and the lower leg circumference was also reduced, resulting in the compression therapy losing effectiveness after a period of time. Therefore, a replacement or rewrapping is needed after bandages show an initial pressure drop after their application, to maintain the efficacy for further compression treatment (Kumar, Das and Alagirusamy 2014; 2013).

A study conducted on edema reduction using bandages found that frequent bandage renewals in a high-pressure range were ineffective and may be counterproductive. Patients who had their bandages applied twice per week showed more effective recovery compared with patients with five times per week of bandage application (Moffat et al. 2012). The study was conducted on 82 patients from seven specialist lymphedema centers in the UK (3 centers) and in the USA (4 centers) (Moffat et al. 2012). Patients were suffering from arm or leg lymphedema. Patients were randomly allocated with two different types of bandaging systems i.e. traditional short-stretch bandage (five times per week) and 3M™ Coban™ 2 (three or five times per week for 19 days) (Moffat et al. 2012). The highest rate of volume reduction for lymphedema was achieved with the Coban™ 2 systems applied twice per week (Moffat et al. 2012). A mean reduction of ~19% in legs and ~11% in arms was achieved with Coban™ 2 system, compared to short-stretch bandages with ~11% and ~8% for legs and arms, respectively (Moffat et al. 2012). The effectiveness of re-application of bandage was higher after two days than after five days, which might be attributed to the stress relaxation property of bandage materials, where the elastic property of bandage fabrics may diminish with time.

The repetitive and continuous movement of the lower leg leads to a change in the bandage structure (Kumar, Das and Alagirusamy 2013). Woven bandage fabrics made from 100% cotton were wrapped around a wooden mannequin, which was newly developed based on lower leg dimensions from five healthy volunteers (30–35 years old). The knee and ankle were chosen as critical parts for the bandage as it stretched and relaxed more during continuous bending (Kumar, Das and Alagirusamy 2013). The movements of these parts of the lower leg might alter the bandage fabric's stress relaxation behavior. Fatigue behavior under cyclic extension was measured at various extension levels, i.e. 47.5–52.5%, 45–55%, 77.5–82.5%, and 77.5–85% to investigate the internal stress of bandage fabrics over a period of repeated extension (i.e. 50, 100, 150, 200, 250, 300, and 1000 cycles) using Zwick/Roell Universal Testing Machine (model-Z050) (Kumar, Das and Alagirusamy 2013). It was found that the stress reduction in specimens was at a faster rate with a higher percentage of extension. The reduction in the stress might be caused by the change in complex viscoelastic properties of the bandage materials, friction between the textile structure, and the nature of deformation (Adeli, Ghareaghaji and Shanbeh 2010).

Different pressures due to compression given by bandages not only depend on the tension applied and the number of overlapped layers (in the case of multilayer bandage system) but also depend on fabric properties (e.g. fabric structure, viscoelastic behavior of fiber or yarn) and the anatomical characteristics (size and shape) of the lower leg (Sikka, Ghosh and Mukhopadhyay 2014, Thomas and Fram 2003). The effect of the extension of bandage fabrics over time creates changes in sub-bandage pressure, perhaps due to a decrease of the limb size after compression therapy and changes in fabric characteristics after extension over a period of time (Sikka, Ghosh and Mukhopadhyay 2014, Farah and Davis 2010). The stress relaxation behavior may deform the bandage fabric materials either during movement or rest, which is caused by a change in the fabric structure when an amount of stress is given to the fabric (Kumar, Das and Alagirusamy 2013). More importantly, the ability of bandage fabrics to provide sustained pressure varies due to fabric structure and varying material type (Kumar, Das and Alagirusamy 2014).

In a different study, patients with heaviness and tension or increasing pain, volume, and edema formation in the lower limb are advised to wear compression stockings (Rabe et al. 2017). According to Rabe et al. (2017), it is suggested that patients with severe congestive cardiac failure use medical compression socking as it may reduce the chances of developing systemic fluid overload. Light compression stockings (15–21 mmHg) are also effective in patients with post-constructive edema after successful bypass surgery (Rabe et al. 2017).

3.6 DISCUSSION

Textile-based compression therapy has been extensively used in managing a range of chronic diseases including edema of the lower extremities, with diverse materials, structural designs, and technology used in the fabrication methods (Kankariya, Laing, and Wilson 2021, Kankariya 2022). Different materials used in compression garments have a significant effect on the physical performance, i.e. fabric structure, yarn, and fibers. For example, fabric with higher density will provide more pressure than that of less density. The effectiveness of the compression therapy not only depends on the materials but also on the skill of the person applying the system and the pressure exerted by the compression device (Rabe et al. 2021, Schreurs et al. 2022).

3.7 CONCLUSION

Since compression textiles are viscoelastic, the interface pressure exerted degrades over time along with the reduction in venous volume and, as a result, a decrease in circumference of the lower leg. The ability of a bandage and compression garment to hold stress depends on several parameters, such as fiber type, yarn structure, and fabric structure. Therefore, compression therapy plays a crucial role in the treatment of lower leg diseases such as edema, deep venous thrombosis, and also venous leg ulcers. Parameters such as yarn or fabric elasticity, layering, and multi-axial stress relaxation of the compression materials are significant in developing a good compression system.

REFERENCES

Adeli, B, Ghareaghaji, A A, and Shanbeh, M. 2010. "Structural evaluation of elastic core-spun yarns and fabrics under tensile fatigue loading." *Textile Research Journal* 81 (2):137–147.

American Society for Testing and Materials. 1980. *ASTM D 3787–80a standard test method for bursting strength of knitted goods – Constant-rate-of-traverse (CRT) ball burst test.* Philadelphia: American Society for Testing and Materials.

Ardakani, T, Asayesh, A, and Jeddi, A A A. 2015. "The influence of two bar warp-knitted structure on the fabric tensile stress relaxation Part II: Locknit, satin, loop-raised." *The Journal of The Textile Institute* 107 (11):1–12.

British Standards Institution. 1968. *Methods of sampling and physical testing of leather – Measurement of distention and strength of grain by the ball burst test.* London: British Standards Institution.

Callam, M J, Haiart, D, and Farouk, M. 1991. "Effect of time and posture on pressure profiles obtained by three different types of compression." *Phlebology* 6:79–84.

Farah, R S, and Davis, M D P. 2010. "Venous leg ulcerations: A treatment update." *Current Treatment Options in Cardiovascular Medicine* 12:101–116.

Geršak, J, Šajn, D, and Bukošek, V. 2005. "A study of the relaxation phenomena in the fabrics containing elastane yarns." *International Journal of Clothing Science and Technology* 17 (3/4):188–199.

Hafner, J, Botonakis, I, and Burg, G. 2000. "A comparison of multilayer bandage systems during rest, exercise, and over 2 days of wear time." *JAMA Dermatology* 136 (7): 857–863.

Hashemi, N, Asayesh, A, Jeddi, A A A, and Ardakani, T. 2016. "The influence of two bar warp-knitted structure on the fabric tensile stress relaxation part I: Reverse locknit, sharkskin, queens' cord." *The Journal of The Textile Institute* 107 (4):512–524.

Kankariya, N. 2022. "Material, structure, and design of textile-based compression devices for managing chronic edema." *Journal of Industrial Textiles* 52:1–35.

Kankariya, N, Laing, R M, and Wilson, C A. 2021. "Textile-based compression therapy in managing chronic oedema: Complex interactions." *Phlebology* 36 (2):100–113.

Kumar, B, Das, A, and Alagirusamy, R. 2013. "An approach to examine dynamic behaviour of medical compression bandage." *The Journal of The Textile Institute* 104 (5):521–529.

Kumar, B, Das, A, and Alagirusamy, R. 2014. "Effect of material and structure of compression bandage on interface pressure variation over time." *Phlebology* 29 (6):376–385.

Kumar, B, Hu, J, and Pan, N. 2016. "Smart medical stocking using memory polymer for chronic venous disorders." *Biomaterials* 75:174–181.

Maqsood, M, Hussain, T, Malik, M H, and Nawab, Y. 2016. "Modelling the effect of elastane linear density, fabric thread density, and weave float on the stretch, recovery, and compression properties of bi-stretch woven fabrics for compression garments." *The Journal of The Textile Institute* 107 (3):307–315.

Maqsood, M, Nawab, Y, Umar, J, Umair, M, and Shaker, K. 2017. "Comparison of compression properties of stretchable knitted fabrics and bi-stretch woven fabrics for compression garments." *The Journal of The Textile Institute* 108 (4):552–527.

Moffat, C J, Franks, P J, Hardy, D, Lewis, M, Parker, V, and Feldman, J L. 2012. "A preliminary randomized controlled study to determine the application frequency of a new lymphoedema bandaging system." *British Journal of Dermatology* 166 (3):624–632.

Mosti, G B, and Mattaliano, V. 2007. "Simultaneous changes of leg circumference and interface pressure under different compression bandages." *European Journal of Vascular and Endovascular Surgery* 33 (4):476–482.

Mosti, G, and Partsch, H. 2013. "Bandages or double stockings for the initial therapy of venous oedema? A randomized, controlled pilot study." *European Journal of Vascular and Endovascular Surgery* 46 (1):142–148.

Mosti, G, Picerni, P, and Partsch, H. 2012. "Compression stockings with moderate pressure are able to reduce chronic leg oedema." *Phlebology* 27 (6):289–296.

Partsch, H. 1991. "Compression therapy of the legs." *Journal of Dermatologic Surgery and Oncology* 17:799–805.

Partsch, H, Partsch, B, and Braunc, W. 2006. "Interface pressure and stiffness of ready made compression stockings: Comparison of in vivo and in vitro measurements. "*Journal of Vascular Surgery* 44 (4):809–814.

Rabe, E, Földi, E, Gerlach, H, Jünger, M, Lulay, G, Miller, A, Protz, K et al. 2021. "Medical compression therapy of the extremities with medical compression stockings (MCS), Phlebological Compression Bandages (PCB), and medical adaptive compression systems (MAC)." *Der Hautarzt* 72:1–14.

Rabe, E, Partsch, H, Hafner, J, Lattimer, C, Mosti, G, Neumann, M, Urbanek, T, Huebner, M, Gaillard, S, and Carpentier, P. 2017. "Indications for medical compression stockings in venous and lymphatic disorders: An evidence-based consensus statement." *Phlebology* 33 (3):163–184.

Ruznan, W S, Laing, R M, Lowe, B J, Wilson, C A, and Jowett, T J. 2021. "Understanding stress-strain behavioral change in fabrics for compression bandaging." *The International Journal of Lower Extremity Wounds* 20 (3):244–250.

Schreurs, R H P, Manuela, A J, Daisy, P, Bruijn-Geraets, D, Hugo, T C, and Arina, J T C. 2022. "A realist evaluation to identify targets to improve the organization of compression therapy for deep venous thrombosis-and chronic venous disease patients." *PloS one* 17 (8): e0272566.

Sikka, M P, Ghosh, S, and Mukhopadhyay, A. 2014. "The structural configuration and stretch property relationship of high stretch bandage fabric." *Fibers and Polymers* 15 (8):1779–1785.

Su, C-L, Maa, M-C, and Yang, H-Y. 2004. "Structure and performance of elastic core-spun yarn." *Textile Research Journal* 74 (7):607–610.

Thomas, S. 2003. "The use of the Laplace equation in the calculation of sub-bandage pressure." *The European Wound Management Association* 3 (1):21–23.

Thomas, S, and Fram, P. 2003. "Laboratory-based evaluation of a compression-bandaging system." *Nursing Times* 99 (40):24–28.

Thomas, W. 1971. "Celanese bagging test for knit fabrics." *Textile Chemist and Colorist* 3 (10):57–59.

Veraart, J C J, Daamen, E, and Neumann, H A M. 1996. "Interface pressure measurements underneath elastic and non-elastic bandages." *Phlebologie* 26:19–24.

Zhang, X, Li, Y, Yeung, K W, and Yao, M. 2000. "Viscoelastic behavior of fibers during woven fabric bagging." *Textile Research Journal* 70 (9):751–757. doi:10.1177/004051750007000901.

Zou, Z Y. 2012. "Study of the stress relaxation property of vortex spun yarn in comparison with air-jet spun yarn and ring spun yarn." *Fibres and Textiles in Eastern Europe* 20 (1):28–32.

4 A Model for Compression Textile Design Based on Dynamic Female Lower Body Measurements

Linsey Griffin, Alison Cloet, and Elizabeth Bye

CONTENTS

4.1 INTRODUCTION

The development of well-fitting, wearable garments rests on a fundamental understanding of the body in motion (LaBat and Ryan 2019). This design principle is critical for compression garments, where a detailed knowledge of changes in shape, form, and size must inform product development. Composed of technical elastomeric fibers and yarns, compression garments are functional, wearable products that

DOI: 10.1201/9781003298526-4

deliver mechanical pressure on the body surface for stabilization, compression, and support of underlying tissues (MacRae et al. 2011). The benefits of compression therapy for the lower body include the reduction of swelling and inflammation (Partsch et al. 2004), supporting wound management (O'Meara et al. 2012), and improving blood circulation (Azirar et al. 2019). It is also beneficial in athletic or aesthetic applications to enhance performance or beautify the body (Xiong and Tao 2018).

Women's compression garments for the lower body are available in forms including briefs, leggings, stockings, knee sleeves, and socks. Garments may apply a universal or graduated pressure to the intended area. In addition to the textile properties, the effectiveness of a compression garment is contingent on the body–product relationship. This is a significant design challenge as the human body is complex, variable, and dynamic, changing in shape, form, and size with considerable variability in skin deformation at major joints. For example, during knee flexion, the skin stretches at the front thigh as the quad muscles lengthen, while the skin shortens at the back thigh as the hamstrings contract. Despite our awareness of the dynamic body, static anthropometry remains the norm in an industry that continues to design products based on body measures in a standing A-frame position. Consequently, traditional methods of measuring the body, such as circumference and length, are limited in their ability to provide dynamic anthropometric data to develop well-fitting compression garments (Granberry et al. 2017, Klepser et al. 2020). Moreover, publicly available anthropometric databases (e.g., ANSUR II, CAESAR) are limited in the diversity of populations such as the aged and lack the critical knowledge of skin deformation and movement.

Advances in 3D scanning technologies provide new methods of discovering dynamic body measurements. Skin deformation research has advanced to understanding dimensional changes of the body from static to flexed or seated positions (Choi and Ashdown 2011, Choi and Hong 2015, Obropta and Newman 2016, Xie and Mok 2022). While these studies underline the implications of skin strain in developing tight-fitting garments, participants have been limited to healthy young adult populations. Little research has examined dynamic measurements of aging women whose body shape, size, and skin elasticity diverge considerably from young adults. Skin deformation research applies dynamic measurements of the lower body to pattern development; however, applications primarily focus on functional sportswear design (Lee et al. 2017, Shi 2020, Wang et al. 2021). Expanded research to examine skin deformation of the aging female body is needed.

This chapter examines the role of dynamic anthropometry, anatomy, and skin deformation in developing female compression garments for the lower body. Drawing on past research, two case studies are presented to illustrate pattern and design considerations for lower body compression garments according to dynamic measurement data.

4.2 ANATOMY OF THE LOWER BODY

The anatomical features of the female body are distinct, so products must consider the bone, muscle, fat, and skin when designing compression products for women.

The pelvic girdle supports the torso of the body on the legs. The sacrum is a shield-shaped bone at the base of the spine that attaches to the pelvis at the sacroiliac (SI) joint (LaBat and Ryan 2019). The hip bones join the sacrum at the SI joints. The right and left hip bones meet at the pubic joint at the front of the pelvic girdle. While the shape of the sacrum can vary, impacting posture and the quality of the SI joints, the shape of the lower spine can tilt the hip, impacting the fit of lower body garments (LaBat and Ryan 2019).

The SI joint requires a balance of mobility and stability to allow for walking while also bearing loads. There are no muscles that stabilize the SI joint directly, but rather a web of muscles from the abdomen. The shape and size of the muscle groups in this area run from the lower torso through the legs and can vary greatly, influencing the size and shape of garments (LaBat and Ryan 2019). Females exhibit fat deposits around the lower abdomen, hips, buttocks, and thighs, and the resulting shape of the body can vary according to individual patterns.

While bones can move at their joints with the help of muscles that stretch and contract, skin covers and supports all body systems. Skin is exceptionally resilient, stretching and recovering regardless of the direction the body moves (LaBat and Ryan 2019). Skin provides the surface that interacts with our clothing and may hold untapped clues about dynamic fit and sizing.

4.3 ANTHROPOMETRY OF THE LOWER BODY

Anthropometry is the science of measuring and describing the dimensions of the human body. Attwood et al. (2004) suggest there are two categories of anthropometric data:

- *Static anthropometry* includes measures of the body in a fixed or resting position, such as height and head circumference.
- *Dynamic anthropometry* includes measures of the body in motion, such as skin deformation and knee flexion and extension.

Mass-produced apparel uses standardized, incremental sizing systems developed using limited static measurements taken from samples that are not representative of the current population diversity (LaBat 2007). Consequently, these systems lack detailed anthropometrics that accurately reflect the body shapes and sizes of the population and assume that basic patterns will accommodate various body shapes within each size category (Ashdown 2007). Though sizing standards and anthropometric surveys have improved, the current system does not adequately support the development of compression garments. These standards are missing critical information regarding the dimensional change of the lower body. Research demonstrates the rate of dimensional change varies considerably from standing to seated positions in different areas and for different body types (Choi and Ashdown 2011, Choi and Hong 2015, Griffin et al. 2019).

Traditionally, anthropometric surveys collected body measurements using manual tools (e.g., anthropometers, measuring tapes, and calipers) and were limited to military populations (Gordon et al. 1989). Since the early 2000s, anthropometric databases expanded to include full-body 3D scans for greater body shape and size analysis. Databases such as ANSUR II and CAESAR featured full-body 3D scans in standing and seated postures for adult military and civilian populations, respectively (Gordon et al. 2014, Robinette et al. 2002). However, measurements were still limited to circumferences and linear breadths and not representative of the change that occurred from standing to seated positions (Granberry et al. 2017). To integrate anthropometric data of body surface changes into compression garment pattern development, focus on dynamic measurements of the lower body that reflect expansion/lengthening and contraction/shortening is needed.

4.3.1 Advances in Dynamic Measurements of the Lower Body

Research utilizing 3D technologies has new methods of apparel design based on dynamic anthropometry. Researchers and designers of compression garments must match everyday users' dynamic range of motion in products that cover the lower torso. In the past, the analysis of the 3D body in dynamic poses has been limited due to poor resolution, lack of color, and obstructed views in critical areas such as the hip and crotch. Scanning technologies are still developing, yet their capabilities supporting the analysis of surface measurements and the body in motion are promising.

A relatively new innovation in 3D body scanning is the capture of full-color models. Recent studies leverage full-color scanners to apply more precise landmarks on the skin with tools such as a washable marker or stamp (Lee et al. 2013, Griffin et al. 2019, Barrios-Muriel et al. 2019). This has advanced dynamic anthropometry by developing new landmarks and measurements for understanding the body in motion. Skin deformation research benefits from color scanning to track how the skin moves and the body changes between different body positions. These findings can be translated into pattern development for functional apparel to improve sizing and fit (Lee et al. 2017, Shi 2020, Wang et al. 2021). However, applications in skin deformation data applied to pattern development for female compression garments of the lower body are not fully explored.

4.4 DYNAMIC ANTHROPOMETRY DATA DESCRIPTION

This dynamic dataset was created to understand how measurements of the aging body change in the seated and standing positions. The landmarking, scanning, and measurement method was developed to enable an in-depth analysis of circumference measurements and shape change, including the expansion and contraction of the lower body in the standing and seated posture. The dataset was applied in two case studies.

4.4.1 Population

This dataset consists of 67 female participants from the US recruited based on size, age, and ethnicity. The average age of the participants is 52.94 years old and ranges from 37 to 75 years old. Overall, 40 Caucasian, 10 Hispanic, nine Black, and eight Asian subjects are represented. The average BMI of the sample is 31.9, and the dataset is equally distributed for participants ranging from normal (18.5–24.9), overweight (25.0–29.9), and obese (>30.0). The Institutional Review Board of the University of Minnesota approved this human subjects research.

4.4.2 Data Description

The body was landmarked prior to scanning and developed to use the scanning system's texture capabilities and to allow accurate tracking of the body between positions. The landmarking method included lines that divided the body into quadrants using horizontal circumferential lines and vertical lines, bony anatomical landmarks, diagonal circumference lines, and natural crease lines such as the horizontal gluteal fold and the groin crease (see Figure 4.1). Each landmark was selected based on the ability to apply the derived dynamic measurements to wearable products. The brief line landmark was prioritized to enable measurement analysis of a product line that diagonally crosses the body.

Washable markers were used to mark the landmarks on the body: red, blue, and green for light skin tones, and metallics of the same colors for dark skin tones. Line placements and bony landmarks were selected from anthropometric definitions by SizeUSA, ISO 7250-1, or CAESAR datasets. A cross-line self-leveling laser was used to ensure level line placement on the body.

The handheld Artec Eva was selected for scanning because it enabled unobstructed views of the positions, had excellent resolution, and captured texture/color in the scan. The star position was chosen to gain better visibility in the hip and thigh region. To standardize this position, the legs of the participant were placed at a 20-degree angle from the top of the thigh in both standing and seated positions. The participants used hiking poles to stabilize themselves in the standing position. A four-foot-tall seat with a plexiglass top was used for the seated position.

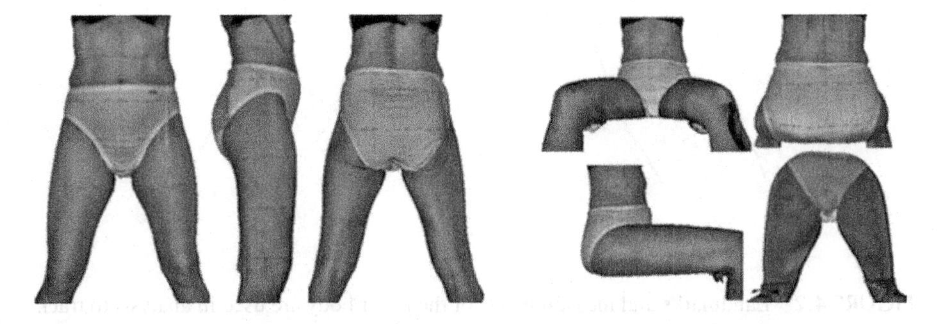

FIGURE 4.1 Example of a scanned participant in standing and seated position with unobstructed views of the waist-hip-thigh region.

The textured scan provided clear, colored visuals of the body landmarks, which were used as cues for digital landmarking. Digital landmarks were placed over the scanned body following the intersection of lined landmarks.

Surface measurements were manually measured from the scans using Anthroscan software. Due to the visible texture, the lines from the landmarks were digitally traced along the body's surface, which significantly increased measurement validity. Distances were captured in millimeters (mm) for circumferences, quadrant surface measures, and vertical length surface measures. Figure 4.2 shows the measurement location on the body.

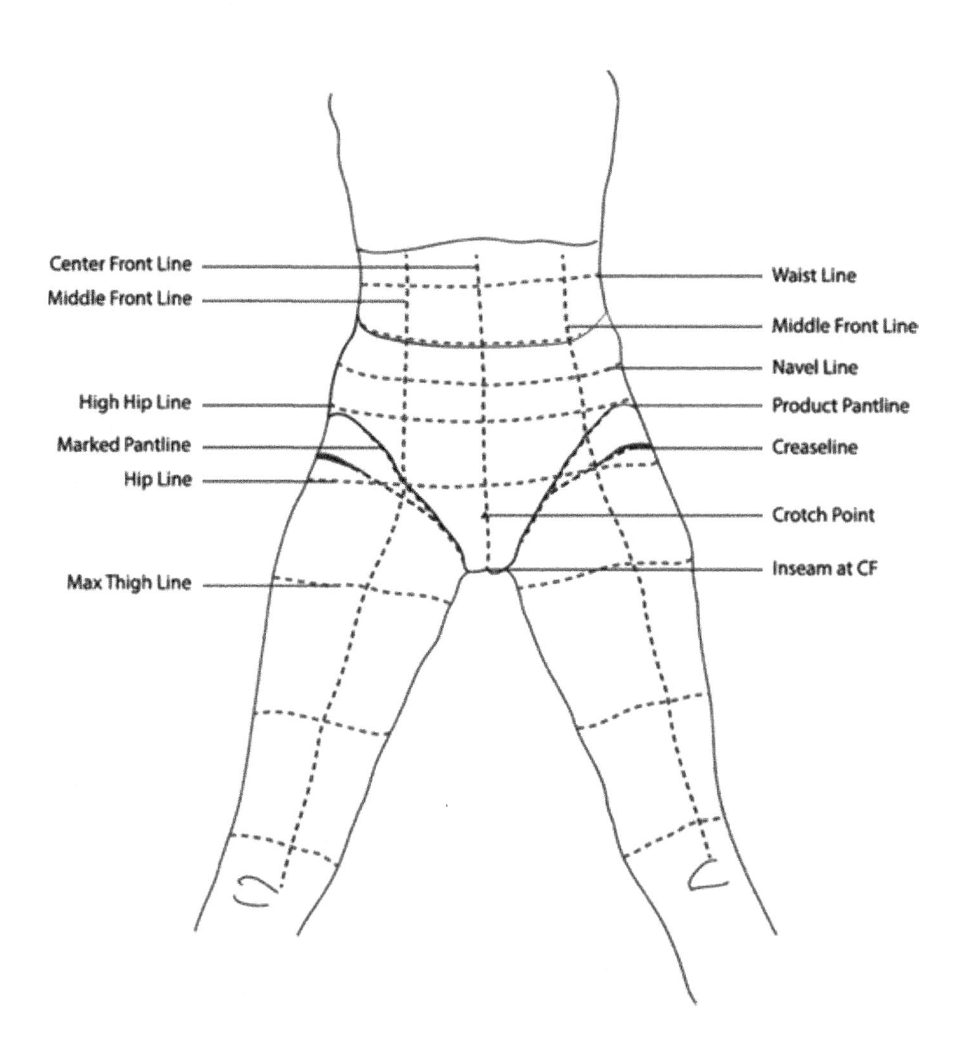

FIGURE 4.2 Landmarks and measurements of the lower body are used in analysis to track dimensional change from standing to seated.

4.4.3 Dynamic Anthropometry Results

This dataset focused on how the aging female body changes in standing and seated positions. Using the quadrant measurement technique, 25 surface measurements were calculated for each subject in both standing and seated positions. The percent change was calculated for each measurement to understand how the body expands and contracts when seated. Paired t-tests were run to determine if the percent change was significant across the dataset for each measurement, and *p*-values were calculated. The results are in Table 4.1.

This dataset captured significant body measurement changes when comparing two body positions. Additionally, expansion and contraction occurred throughout the dataset, especially in crossbody measurements such as the brief line in length measurements (front contraction, back expansion). The hip shows overall expansion from standing to seated, whereas the diagonal circumference of the brief line displayed both contraction and expansion in the quadrant measurements. The rise measurement was shortened in the front and lengthened in the back. The length measurements at the midline front and midline back exhibited some of the most dynamic changes in the dataset.

4.5 BODY–PRODUCT RELATIONSHIP DESIGN MODELS AND APPLICATION

The dynamic anthropometry dataset is applied to two case studies, a compression brief and a compression short, to demonstrate how the percent change data influences 2D pattern shape and garment design.

4.5.1 Case Study One: Application of Anthropometric Data to 2D Pattern Design to a Compression Brief

A women's postoperative (post-op) compression brief (Figure 4.3) is selected for Case Study One. This type of compression garment is worn in the waist-crotch area to support the abdomen following procedures such as C-section, hysterectomy, and other lower abdominal surgeries to speed up recovery and reduce pain, tenderness, and swelling.

The fit of a post-op compression brief must provide full support to the abdominal wall that moves with the body. The waistband and the brief line must not dig into tender tissue or muscles nor be too loose to be ineffective. The medical-grade front compression panel must accommodate the body in expansion and contraction while delivering a consistent pressure distribution. In sum, the garment must provide a comfortable and functional close-fit for every user. This proves a challenge when patterns for post-op compression briefs lack an understanding of the dynamic body and utilize static measures of the waist, hips, and crotch depth. An application of dynamic measures to 2D pattern development demonstrates how the fit of post-op compression briefs can be improved in this case study.

TABLE 4.1

Summary Statistics for Measurement Percent Change from Standing to Seated Position of the Female Lower body

		Standing to seated measurement comparison and change (n = 67)			
Measurements		**Mean % change**	**Dynamic description**	***t*-Test**	***p*-Value**
Waist	Total circumference	0.06	Slight expansion	−.439	.662
	CF to MF	3.56	Slight expansion	−4.306	.000***
	MF to SS	14.87	Moderate expansion	−1.347	.183
	SS to MB	−2.82	Slight contraction	4.235	.000***
	MB to CB	−2.31	Slight contraction	1.416	.162
Hip	Total circumference	7.90	Slight expansion	−10.522	.000***
	CF to MF	8.79	Slight expansion	−7.748	.000***
	MF to SS	11.03	Moderate expansion	−6.989	.000***
	SS to MB	9.23	Slight expansion	−6.143	.000***
	MB to CB	5.89	Slight expansion	−7.930	.000***
Brief line (diagonal circumference across body)	Total circumference	7.13	Slight expansion/lengthening	−4.756	.000***
	Inseam to MF	−23.77	Substantial contraction/shortening	5.861	.000***
	MF to SS	2.45	Slight expansion/lengthening	−1.869	.075
	SS to MB	1.61	Slight expansion/lengthening	−2.091	.042
	MB to inseam	27.08	Substantial expansion/lengthening	−9.720	.000***
Max thigh	Total circumference	11.58	Moderate expansion	−19.978	.000***
	Inseam to MF	24.26	Substantial expansion	−11.179	.000***
	MF to SS	13.15	Moderate expansion	−15.960	.000***
	SS to MB	−2.65	Slight contraction	1.547	.127
	MB to inseam	5.62	Slight expansion	−1.970	.055

(*Continued*)

TABLE 4.1 CONTINUED

Summary Statistics for Measurement Percent Change from Standing to Seated Position of the Female Lower body

		Standing to seated measurement comparison and change ($n = 67$)			
Measurements		**Mean % change**	**Dynamic description**	***t*-Test**	***p*-Value**
Length measurements	Total rise	–5.64	Slight shortening	12.632	.000***
	Rise front: CF to inseam	–30.83	Substantial shortening	25.551	.000***
	Rise back: CB to inseam	12.99	Moderate lengthening	–12.044	.000***
	MF: Waist to creaseline	–31.67	Substantial shortening	15.574	.000***
	MB: Waist to creaseline	22.51	Substantial lengthening	–24.393	.000**

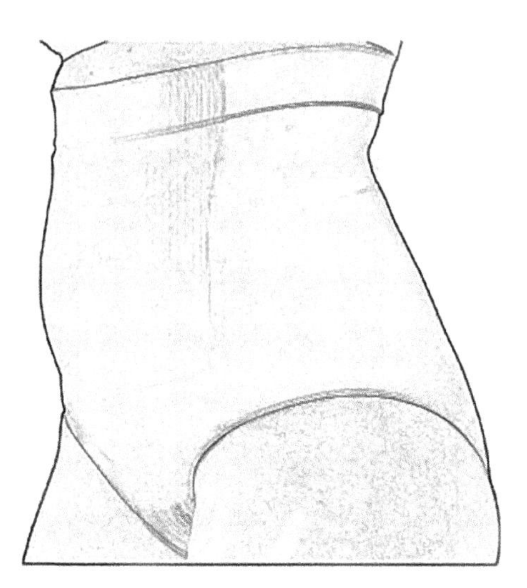

FIGURE 4.3 Example of compression brief for Case Study One.

4.5.1.1 2D Pattern Development

The base compression brief pattern was created to represent how traditional patterns are created: using only standing/static measurements with minimal curved shaping. The pattern covered the waist, hip, and crotch region with a leg opening that skimmed the groin and gluteal creases. To prepare the base pattern to be dynamically altered, the pattern parts were labeled and landmarks added to match the dataset (i.e., midline front and back were drawn on the pattern).

A slash and spread technique was used to alter the base pattern, and the percent change data was applied. The resulting 2D pattern represents the body change that occurred between the static standing position and the dynamic seated position.

4.5.1.2 Dynamic Pattern Shape Comparison

The static and dynamic pattern shape development represents a method to apply dynamic anthropometric data to compression garment design. Often, designers and engineers rely heavily on textile structures and do not consider the body shape when creating the initial patterns. When the basic garment shape does not match the body, the result is often uneven compression and ill-fit.

The patterns seen in Figures 4.4 and 4.5 represent the body in both standing and seated positions. The comparisons in Figures 4.4 and 4.5 illustrate that the front contraction and shrinking of dimensions result in a significant shape change. The front rise compresses and protrudes at the center front, while the brief line area compresses and shrinks. Additionally, the pattern begins with straight lines at the center front/back, waist, and side seam. The dynamic data transforms the basic pattern shape to include curvilinear lines throughout the pattern, representing the actual curvature change exhibited by the body. The pattern back shows a more

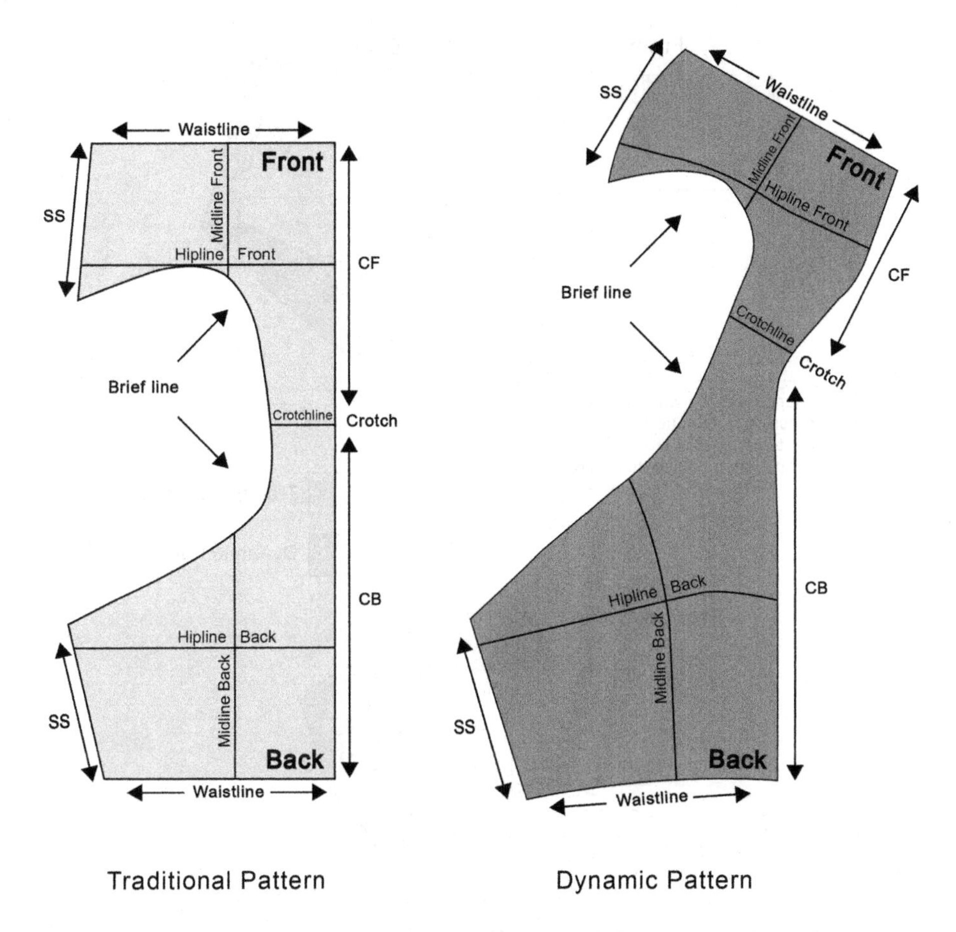

FIGURE 4.4 Comparison of traditional compression brief pattern and dynamic compression brief pattern.

even expansion in the hip, while the back brief line creates a significant elongation across the body. Based on the pattern development presented for this case study, there are substantial implications for 2D pattern shape, knit structure, and grainline placement.

Dynamic, dimensional shape changes need to be incorporated into the 2D pattern for compression briefs to improve product performance and fit. The base pattern for each size should match the body shape and dimensional change of that user group. Traditional grade rules may not apply to sizing these products because of the diverse shape/dimensional change variations seen among different-sized females. The development of sizing systems and grade rules will be significantly influenced by dynamic anthropometric data, especially how bodies are categorized and how specific shape change is incorporated into the base pattern for each size. Also notable is the transformation along the brief line's diagonal circumference. The implications for product

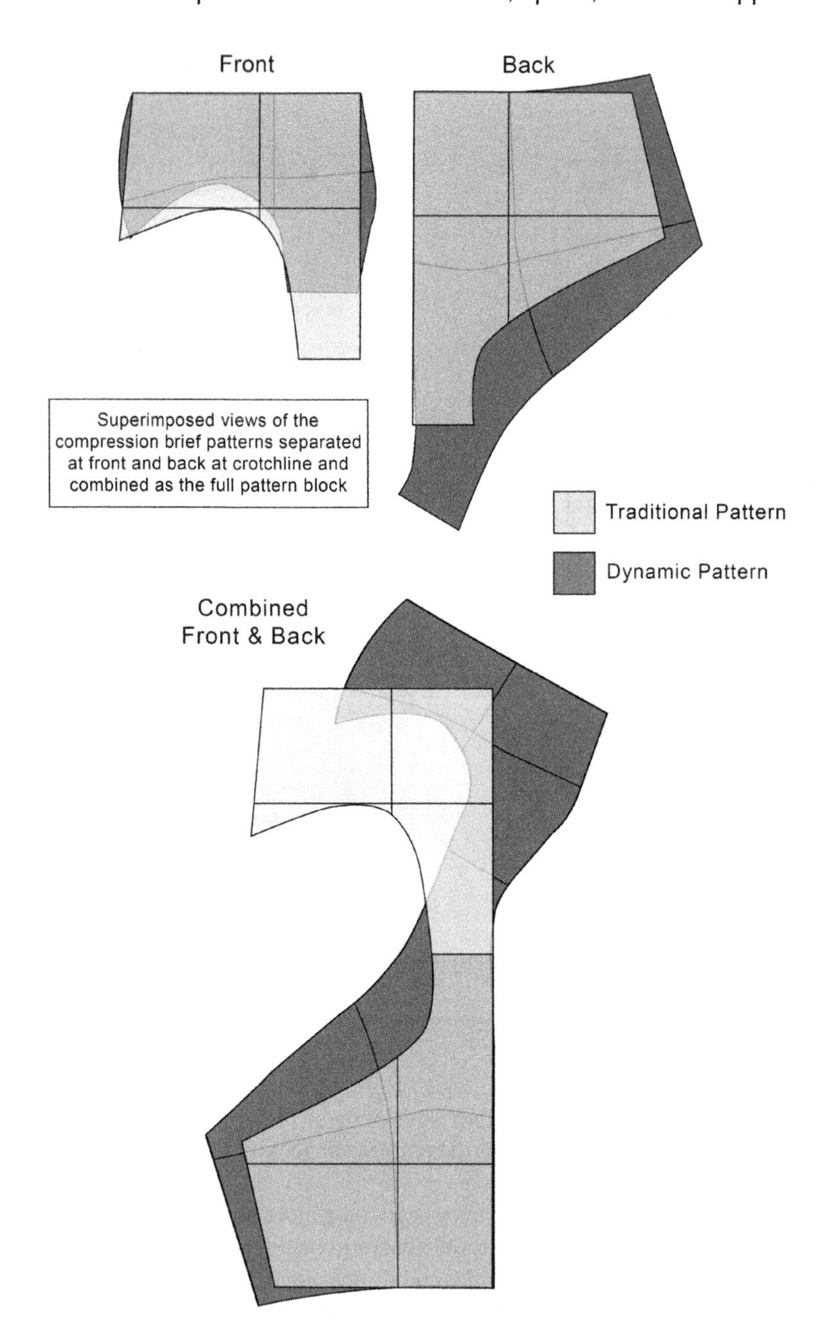

FIGURE 4.5 Superimposed view of traditional and dynamic compression brief pattern differences.

design and style line placement based on diagonal circumference data is significant, especially as functional requirements of medical compression products increase.

As designers and engineers integrate dynamic body data into compression garments, it is imperative to consider internal body anatomy as organs/muscles/ligaments shift and reposition with body movement. Understanding internal body anatomy, combined with dimensional change data, will enable the development of more sophisticated medical compression garments.

4.5.2 Case Study Two: 3D Application of Data to Knit Structure Design on Compression Therapy Shorts

Case Study Two reviews women's compression therapy shorts (Figure 4.6). Compression shorts are designed to improve circulation and provide targeted joint support in the waist-hip-thigh area. These shorts can be worn to manage conditions such as lymphedema, varicose veins, and muscle strains and provide maternity and postpartum support or speed recovery following lower abdomen surgeries.

The fit of compression shorts must provide full support to the abdomen, hips, and thighs while delivering a functional, comfortable amount of pressure that accommodates the static and dynamic body. All components of the shorts, including the waistband and thigh sleeve, must provide a close fit for every user. Traditionally, compression shorts are made using a synthetic-elastane fiber blend that applies either a universal or graduated percentage of compression throughout the leg. However, the body requires varying amounts of compression in the waist-hip-thigh area.

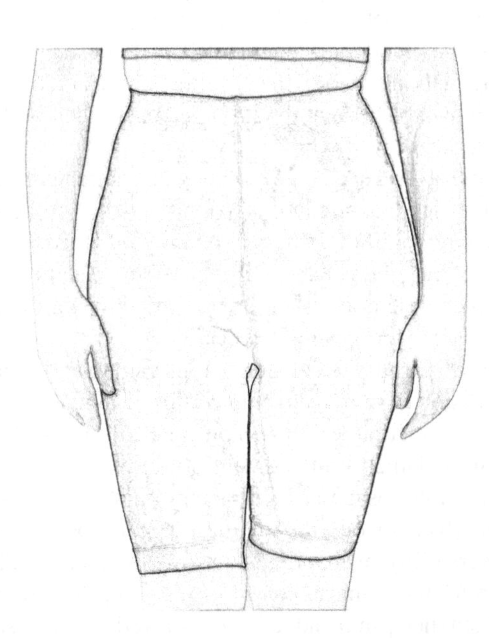

FIGURE 4.6 Example of compression therapy shorts for Case Study Two.

Consequently, it is necessary to integrate various knitwear structures into compression shorts design to support the lower body's natural expansion and contraction. This case study considers the placement of different compression knitwear textiles linked to knowledge of the dynamic body in 2D pattern development for compression shorts.

4.5.2.1 Garment Construction and Knit Design

The construction of a well-fitting lower-body compression garment is complex. While knitted structures have multi-directional extensions to conform to the body, negative ease in the pattern must be carefully calculated to ensure the garment fits like a second skin. This is further complicated by the variety of stress-strain behaviors of compression textiles, which must be skillfully placed in pattern development to align with the body's support needs. A tight fit may risk discomfort or adverse physiological effects (Chan and Fan 2002), while a loose fit compromises support and can reduce thermal comfort due to air gaps between the garment and the body (Klepser et al. 2020).

Figure 4.7 displays the average dimensional change from standing to seated positions of 67 female participants as applied to the construction and knit design of a compression short. Horizontal lines of the body correspond to the course direction of a knit, while vertical lines of the body correspond to the wale direction. Diagonal lines represent both directions. In the course direction, the body substantially expands at the inner front of the thigh. Moderate expansion occurs along the sides of the front waist, hip, and thigh lines. Meanwhile, skin slightly contracts at the back waistline and back outer max thigh line, slightly expanding at the back hipline. In the wale direction, the center front rise and mid front rise exhibit substantial shortening. The center back rise has moderate lengthening and substantial lengthening at the mid back. In the diagonal direction of the brief line, there is substantial contraction at the center crotch, slight expansion at the front and back sides, and substantial expansion at the center buttocks.

After mapping the average change in body measurements in Figure 4.6, the dynamic anthropometric data are applied to the design of compression shorts. Figure 4.8 presents a model illustrating the dynamic placement of compression knit textiles on the female lower body. The model does not specify the intensity of compression dosage; rather it demonstrates the alignment of knit textile behavior with skin deformation behavior for apparel production.

The findings of Case Study Two argue compression shorts require a segmented and gradient compression design to accommodate the dynamic body. This is a complex request in mass production. Compression garments are generally manufactured via circular knitting or flatbed knitting with a universal or gradient knit structure that changes compression dosage at incremental circumferences (Liu et al. 2017). Circular knitting is advantageous for seamless apparel design, while flatbed knitting offers greater versatility in shaping and sizing the pattern. Flatbed knitting is a more feasible option for the compression shorts design proposed in the model where needles, yarns, and stitches can be added and removed to adjust shape and compression during manufacturing.

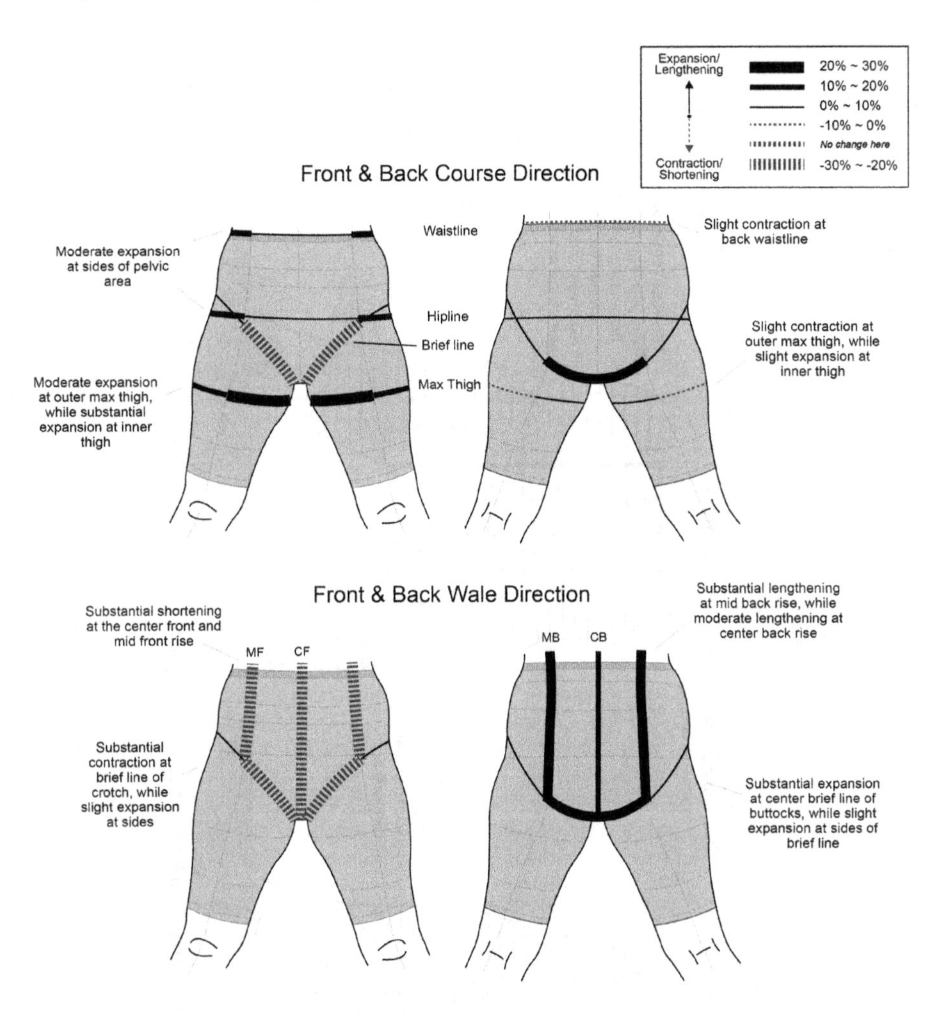

FIGURE 4.7 Average lower body dimensional change from standing to seated position ($n = 67$).

Figure 4.8 describes the behavior of knit structures at a theoretical level of elasticity and stiffness, which are vital characteristics that affect pressure performance, efficacy, comfort, and durability (Liu et al. 2017). While compression shorts are unlikely to include a diagonal circumferential style line, such as the brief line, the dynamic change near the gluteal and groin crease lines and diagonally across the body requires complex integration of the mechanical properties of compression textiles. This area needs a transitional knit structure with length-, diagonal-, and widthwise stretch to accommodate multi-directional expansion and contraction. By contrast, the front pelvic region requires a stable knit with minimal stretch to counter expansion and allow for contraction in this area. To integrate a stiffer compression fabric, appropriate knit structures may utilize an inelastic or short-stretch inlay yarn to minimize widthwise stretch.

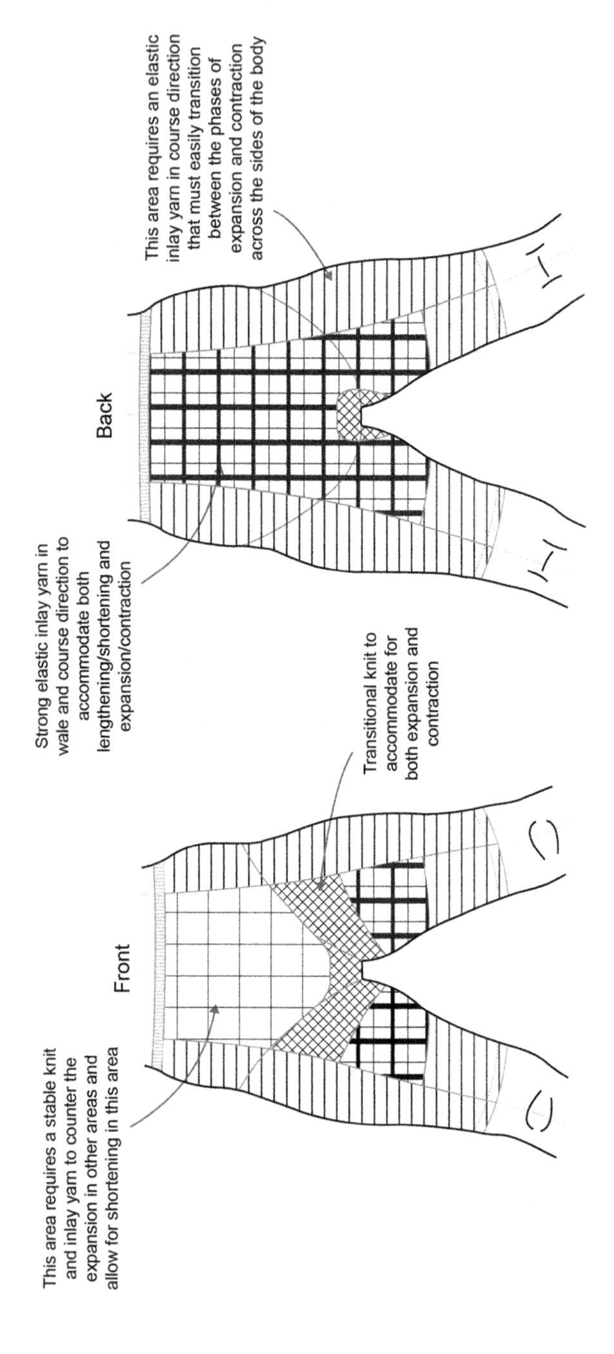

FIGURE 4.8 Model for dynamic placement of compression knit textiles on the female lower body.

Across the body's front and back sides and lower thigh, a knit structure must dynamically shift between phases of expansion and contraction. Appropriate knit structures may carry an elastic inlay yarn with long stretch properties and high recovery to accommodate expansion and relaxation behaviors. Finally, the center buttocks and inner thigh area experience substantial expansion/lengthening, which requires a knit structure with long stretch properties in length- and widthwise directions. Inlay yarns may utilize a high elastane core or incorporate elastic fibers with high extensibility and recovery.

4.6 CONCLUSION AND FUTURE DIRECTIONS

The complex interaction of the body–product relationship can be improved through a deep understanding of the dynamic body. These case studies demonstrate the potential for enhanced functional design for compression briefs and shorts through the application of dynamic anthropometric data. Transforming 2D flat pattern design and creating effective dynamic knit structural capacity to match the body during movement is feasible with careful consideration of body anatomy and dynamic anthropometry. For optimal fit and function, compression garment design should include a holistic system of pattern shape, ease, size grading, knit textile development, garment construction, and product performance integrated with knowledge of the dynamic body.

The application of this dynamic dataset demonstrated the importance of incorporating body shape and ease into the 2D flat pattern shape to represent how the body changes while moving: expansion, contraction, lengthening, and shortening across vertical, horizontal, and diagonal orientations. Integrating pattern shapes into compression garment design is a departure from traditional knit compression garments, especially for those created on circular knitting machines. Shape integration in the design of the 3D form may require significant rethinking and adjustments to the traditional manufacturing process. Regardless, compression products that reflect the true shape of the body in all positions will serve the wearer at the most optimal level of comfort and performance.

The most dynamic area of the lower body examined is along the diagonal circumference of the brief line that surrounds the gluteal and groin crease lines. When designing garments, there is a tendency to only consider traditional horizontal circumferences and vertical lengths. However, the changes that occur along the horizontal and vertical planes of the body are magnified along some diagonal, crossbody lines. The implications of this diagonal movement include consideration of how knit structures are oriented on the body, the placement of seams, design lines, and functional features. The process for accommodating shape into both the flat pattern and mechanical knit structures must be adapted. Future research testing the complex interaction of these elements on the dynamic body is needed to advance compression garment design.

There is a tremendous opportunity for 3D knit technology to be integrated into compression garments as a way to bridge complex data into new product design. Creating advanced 3D knit structures, as well as integrating cutting-edge materials such as shape memory alloys into compression products, enables the innovative application of dynamic body data to reflect the true shape of the body. Current functional compression deficiencies may be solved by advancing 3D knit structure research to reflect the dynamic body.

The impact of 3D and 4D data on design is unlimited. However, the wrong data can misinform the product shape, design features, fit, function, and safety. During the process of creating relevant data for the design of compression products, it is imperative that the body–product relationship is defined. The decisions made at the method level, including landmarking, body stabilization, and measurement extraction, influence how the body data is integrated into the design. The dataset used in these case studies was developed specifically for the application to products that fit close to the body. Future development of dynamic body data will need to challenge existing 3D/4D processes and standards to enable accurate, relevant, and innovative data integration into design.

The future of compression garment innovation will involve diverse disciplines working together to strengthen core knowledge of the dynamic body. The holistic system of compression products will need to lean on interdisciplinary collaboration to leverage data integration of dynamic anthropometry into the mechanical knit structure and design of compression garments.

REFERENCES

Ashdown, S P. 2007. *Sizing in clothing: Developing effective sizing systems for ready-to-wear clothing*. Cambridge: Woodhead Publishing Limited.

Attwood, D A, Joseph M D, and Mary, E D. 2004. "Personal factors." In *Ergonomic solutions for the process industries*, edited by Attwood, D A, Joseph, M D, and Mary, E D, 29–63. Burlington: Gulf Professional Publishing. doi: 10.1016/B978-075067704-2/50004-0.

Azirar, S, Diebrecht, A, Martin, H P, Martino, H N, Adriaan, N P F, and Dinanda, N K. 2019. "Compression therapy for treating post-thrombotic syndrome." *Cochrane Database of Systematic Reviews* 2019 (9): 1–33. John Wiley & Sons, Ltd. doi: 10.1002/14651858.CD004177.PUB2.

Barrios-Muriel, J, Sánchez, F R, Sánchez, F J A, and Salgado, D R. 2019. "In vivo measurement of surface skin strain during human gait to improve the design of rehabilitation devices." *Computer Methods in Biomechanics and Biomedical Engineering* 22 (15): 1219–1228. doi: 10.1080/10255842.2019.1655549.

Chan, A P, and Fan, J. 2002. "Effect of clothing pressure on the tightness sensation of girdles." *International Journal of Clothing Science and Technology* 14 (2): 100–110. doi: 10.1108/09556220210424215.

Choi, S, and Ashdown, S P. 2011. "3D body scan analysis of dimensional change in lower body measurements for active body positions." *Textile Research Journal* 81 (1): 81–93. doi: 10.1177/0040517510377822.

Choi, J, and Hong, K. 2015. "3D skin length deformation of lower body during knee joint flexion for the practical application of functional sportswear." *Applied Ergonomics* 48: 186–201. doi: 10.1016/j.apergo.2014.11.016.

Gordon, C C, Churchill, T, Clauser, C E, Bradtmiller, B, and McConville, J T. 1989. "1988 anthropometric survey of U.S. army personnel: Methods and summary statistics." Yellow Springs, OH: Anthropology Research Project, Inc. https://apps.dtic.mil/sti/pdfs/ADA225094.pdf

Gordon, C C, Blackwell, C L, Bradtmiller, B, Parham, J L, Barrientos, P, Paquette, S P, Corner, B D, Carson, J M, Venezia, J C, Rockwell, B C et al. 2014. "2012 anthropometric survey of U.S. army personnel: Methods and summary statistics." Natick, MA: US Army Natick Soldier Research, Development and Engineering Center. https://apps.dtic.mil/sti/pdfs/ADA611869.pdf

Granberry, R, Duvall, J, Dunne, L E, and Holschuh, B. 2017. "An analysis of anthropometric geometric variability of the lower leg for the fit & function of advanced functional garments." In ISWC '17: Proceedings of the 2017 ACM International Symposium on Wearable Computers, 10–17. doi: 10.1145/3123021.3123034.

Griffin, L, Juhnke, B, Emily, S, Colleen, P, and Karolina, D. 2019. "Method to capture and analyze the Waist-Hip-Thigh body region of seated-standing 3D scans." In 10th International Conference and Exhibition on 3D Body Scanning and Processing Technologies, 254–265. doi: 10.15221/19.254.

Klepser, A, Morlock, S, Loercher, C, and Schenk, A. 2020. "Functional measurements and mobility restriction (from 3D to 4D scanning)." In *Anthropometry, apparel sizing and design* (Second edition), edited by Zakaria, N and Gupta, D, 169–199. Woodhead Publishing. doi: 10.1016/B978-0-08-102604-5.00007-X.

LaBat, K L. (2007). "Sizing standardization." In *Sizing in Clothing: Developing Effective Sizing Systems for Ready-to-Wear Clothing*, edited by Ashdown, S P, 88–107. Cambridge: Woodhead Publishing.

Labat, K L, and Ryan, K S,. 2019. *Human body: A wearable product designer's guide*. Boca Raton, FL: CRC Press.

Lee, H, Hong, K, and Lee, Y. 2013. "Ergonomic mapping of skin deformation in dynamic postures to provide fundamental data for functional design lines of outdoor pants." *Fibers and Polymers* 14 (12): 2197–2201. doi: 10.1007/s12221-013-2197-6.

Lee, H, Hong, K, and Lee, Y. 2017. "Development of 3D patterns for functional outdoor pants based on skin length deformation during movement." *International Journal of Clothing Science and Technology* 29 (2): 148–165. doi: 10.1108/IJCST-08-2016-0090.

Liu, R, Xia, G, Terence, T L, and Trevor, L. 2017. "A critical review on compression textiles for compression therapy: Textile-Based compression interventions for chronic venous insufficiency." *Textile Research Journal* 87 (9): 1121–1141. doi: 10.1177/0040517516646041.

MacRae, B A, James, D C, and Laing, R M. 2011. "Compression garments and exercise." *Sports Medicine* 41 (10): 815–843. doi: 10.2165/11591420-000000000-00000.

Obropta, E W, and Newman, D J. 2016. "Skin strain fields at the shoulder joint for mechanical counter pressure space suit development." In 2016 IEEE Aerospace Conference, 1–9. doi: 10.1109/AERO.2016.7500744.

O'Meara, S, Nicky, C, Nelson, E A, and Dumville, J C. 2012. "Compression for venous leg ulcers." *Cochrane Database of Systematic Reviews* 2012 (11). doi: 10.1002/14651858.CD000265.PUB3/MEDIA/CDSR/CD000265/IMAGE_N/NCD000265-CMP-023-01.PNG.

Partsch, H, Johann, W, and Bertrand, L. 2004. "Compression stockings reduce occupational leg swelling." *Dermatologic Surgery* 30 (5): 737–743. doi: 10.1111/j.1524-4725.2004.30204.x.

Robinette, K M, Sherri, B, Hein, D, Mark, B, and Fleming, S. 2002. "Civilian American and European surface anthropometry resource (CAESAR), Final Report." Volume 1. Summary. Dayton, OH: Sytronics Inc. https://apps.dtic.mil/sti/citations/ADA406704.

Shi, Q. 2020. *Compression cycling garment design for performance enhancement and muscle fatigue recovery*. Ph.D. diss., The Hong Kong Polytechnic University.

Wang, W, Honglian, C, Zhijia, D, and Zhe, G. 2021. "Digital design model for weft-knitted seamless yoga pants based on skin deformation." *Journal of Engineered Fibers and Fabrics* 16 (February): 1–9. doi: 10.1177/1558925021990503.

Xie, N, and Mok, P Y. 2022. "Investigation of full body skin surface variations under dynamic poses." *International Journal of Industrial Ergonomics* 87: 103237.

Xiong, Y, and Xiaoming, T. 2018. "Compression garments for medical therapy and sports." *Polymers* 10 (6): 1–19. doi: 10.3390/polym10060663.

5 Compression Clothing and Body Deformations Through High-speed 4D Scanning

Tatjana Spahiu and Yordan Kyosev

CONTENTS

5.1 INTRODUCTION

Clothing plays an important role in our lives for different purposes. The prime one is functional to protect our body from different climate conditions and by adding functionality to improve our life. According to their use or purpose, they can be classified as daily wear, sports wear, protective clothing, etc. Within the wider group of clothing, compression cloths are special garments worn on different body parts. As elastic items, they exert mechanical pressure on the body surface for specific purposes.

Compression garments have gained widespread use and are expected to grow. The forecast by the Compression Wear and Shapewear Global Market Report 2022 Market Size predicted that the global shapewear and compression wear market is expected to grow to $5.75 billion in 2026 at a compound annual growth rate (CAGR) of 6.8% (Global Compression Wear and Shapewear Market Trends, Strategies, Opportunities For 2022-2031 2022). According to the report, urbanization is the key factor for this growth. Different characteristics such as properties of materials, fabrics, and the model that includes seam placement can decide the compression properties of the garment. Studies conducted by various researchers depict the impact on

DOI: 10.1201/9781003298526-5

body performance during their use. During intense running, wearing sports compression socks results in improved subsequent running performance (Brophy-Williams, et al. 2018). The physical and mechanical properties of the fabric for compression clothing are of high importance in order to estimate and predict the required force exerted on body parts (Wang, Felder and Cai 2011). The structure of this chapter is organized as follows: an overview of 3D technology for evaluating compression garments is included. Then, a methodology to use 4D scanning for the human body with and without compression socks in movement is included. The main results and discussion are presented, and conclusions are depicted at the end of this work.

The aim of the research presented here will be focused on the methodology to implement a 4D scanning system for taking 3D information of the body surface in movement. Furthermore, advanced software for data manipulation used to create virtual replicas of the human body will be presented. A key study to evaluate the effect of compression socks will be included. Results and discussion of data comparisons between virtual models with and without compression socks will be presented. In the end, the conclusion part is depicted.

5.2 3D TECHNOLOGY FOR EVALUATING COMPRESSION GARMENTS

3D technology has been widely used over the years for different purposes in the fashion industry. During the data capture process, short time, contactless and accuracy of data captured are some of the advantages of these applications, where for garment construction and anthropometric purpose shows suitable reliability (Parker, Gill and Hayes, 3D Body Scanning has Suitable Reliability: An Anthropometric Investigation for Garment Construction 2017). One of the biggest challenges, according to researchers, is predicting pressure variation during wearing. So predicting pressure criteria for specific body parts and real-time display seems to be ongoing research (Xiong and Tao 2018). Also, other authors suggest that wearing compression clothing may reduce muscle pain, damage, and inflammation during recovery (Engel, et al. 2016). Advancements in 3D technology like scanning and virtual simulation for compression clothing have seen wide applications, where various researches are conducted.

Brubacher et al. (2021) report the use of 3D scanning and virtual simulation to compare real and virtual results of numerical pressure values for a small group of participants. Their results depict some limitations of virtual simulation for developing these products as no good relationship was found between virtual and real whole-body compression software. Using a VITUS XXL 3D body scanner, other authors have scanned a group of subjects and evaluated body circumferences by suggesting that body susceptibility should be taken into account the value of unit pressure (Ilska, et al. 2017). One of the main problems with garment fit is not only accurate body measurements but including body shape. This is more sensitive to the process of producing compression socks. In order to receive the maximum effect indeed for this type of product, it is suggested to include not only accurate leg measurements but even the leg shapes which can be generated from 3D body scanning. A clinical study conducted for a small group reveals the application of 3D scanning for the assessment and production of bespoke hosiery by highlighting the positives of bespoke scanning and fitting (Stephen-Haynes and Toner 2012). Combining 3D body scan data, types of fabric, and pressure is investigated by researchers and a method for developing personalized compression garments not only according to body measurements but including a different pressure value exerted on body parts is presented (Salleh, et al. 2012). Including numerical simulation to 3D body scanning and properties of the fabric, researchers have created a numerical model that includes even the dynamic mechanical interactions between fabric and skin during wear (Li, et al. 2019).

Based on this, authors have created a 3D scanning system where scan data users can be used for customized stockings which offer custom pressure on different parts of the legs based on individual data (Zhi, Tingyu and Conghui2015). The use of 3D body scans can ensure the production of compression garments based on individual body measurements that can exert pressure within and below clinical standards (Ashby, et al. 2021). Meanwhile, the case of individual scans from a group of participants and their statistical analysis is used by researchers to create a parametric model of the calf. It can be used for the evaluation or size customization of compression garments (Xi, et al. 2020). In another case, evaluating changes in compression before and after running a seamless knitted sports bra showed a decrease in overall compression. Analyzing 3D body models taken from NX-16 3D body scanner $[TC]^2$ revealed again the usefulness of 3D technology (Gorea and Baytar 2020). 3D data taken from the human body by using a handheld body scanner can be used to develop 3D pressure garments. Results show the effectiveness of this method in achieving the intended pressure that the garment can exert (Salleh, Fozi and Lamsali2017). Modeling patterns of garments from 3D to 2D is another way to use CAD technologies for garment production, such as compression garments. So, transforming 3D body data is used to create a soft virtual model over which a compression garment is modeled and predicts the pressure with higher accuracy (Kuzmichev, Tislenko and Adolphe 2018). It is obvious the accuracy and repeatability to take body measurements from 3D scan data by avoiding contact, as in the case of compression bands (Karpiuk et al. 2019) and the automatic creation of compression clothing may be a good alternative, including the economic advantages too (Kisch, et al. 2021).

5.3 METHODOLOGY FOR 4D BODY SCANNING AND EVALUATING COMPRESSION SOCKS THROUGH VIRTUAL MODELS

A high-speed 4D body scanner has been implemented to take 3D human body scanning during movement from Instituto Biomecanica de Valencia. It is a modular scanning system based on photogrammetry. MOVE 4D scanning is equipped with 12 synchronized 4D scanning modulus and processing software. Each module is composed of three video cameras and an infrared projector that enable us to obtain 3D

FIGURE 5.1 4D scanning system MOVE4D at the ITM of TU Dresden.

point cloud and texture simultaneously for all frames. Figure 5.1 presents a picture taken from the scanning system installed at the Institute of Textile Machinery and High Performance Materials. Table 5.1 depicts the main characteristics of this scanning system.

Before starting the scanning process, a new calibration was carried out through a three-step process. The first step is wand calibration using a wand tool, through the process of sweeping the scanning volume. The second step includes calibration of the coordinate system, and the third step includes global calibration where the automatic algorithm calibrates all the modules globally.

A 58-year-old participant took part in the scanning process. She uses compression socks recommended by the doctor. Compression socks were from Mediven produced in Germany, CCL 2, Size 2, regular with closed toe, composition is 61% polyamide and 39% elastan. According to the standard compression Class II, the compression intensity is medium with a value of 3.1–4.3 kPa (1kPa = 7.5 mmHg) or 23–32 mmHg (1 mmHg = 0.133 kPa). At first, the participant was scanned without compression socks at a normal state. The second scanning process was done by following the same procedure. The difference in leg measurements due to compression socks is evaluated.

Virtual leg models were superimposed in Geomagic Qualify version 2012. A best fit-alignment function which moves the test model, even called floating, in order to

TABLE 5.1
Main Characteristics of MOVE 4D Body Scanner

Characteristics	Values
Dynamic scanning volume	2 m × 3 m × 3 m
Resolution	1 mm (height resolution)
Capture frequency	Up to 180 fps
Scanning time	1 msec/scan
External synchronism	Trigger and synchro input/output
Capture	Synchronized 3D data and texture
Lighting	Inbuilt lighting system
Outputs	Sequence of watertight mesh (OBJ/PLY) with a density of 50,000 points including texture Point cloud (PLY) with a density of more than 50,000 points Set of 93 body measurements

share the same physical space with the reference model is executed. The color-coded map shows the deviations of the test model from the reference model. The differences between girth measurements of the right and left legs with and without socks were calculated. Girth values are calculated at different leg heights where some of the section plans to calculate girth contours are taken according to the standard DIN 58133:2008-07.

5.4 RESULTS AND DISCUSSIONS

Data generated from MOVE 4D scanning process are cloud points aligned by each module. The software part of MOVE 4D scanning system has processing tools that automatically generate the 3D mesh of the human body with homologues correspondence along the sequence of scans. After performing 4D human body scanning, the next step included for further analysis is data export where captures can be exported as PLY and OBJ. The difference between these files here is that they are exported as PLY as a point cloud and OBJ as a watertight mesh of 50,000 points. Figure 5.2 depicts 3D leg models of the participant wearing compression socks.

Meanwhile Figure 5.3 depicts point clouds of 3D legs models taken from a 4D body scanner of the participant with and without socks.

CAD models of participants' legs are presented in Figure 5.4.

Data comparison from 3D leg models of the participants with and without compression socks is given in Figure 5.5. The deviation analysis conducted generates a three-dimensional color-coded mapping between both leg files. The test and reference are superimposed, and some of the data generated from deviation analysis are maximum distance, average distance, and standard deviation.

The maximum distance indicates the greatest deviations that are found from the reference and test leg models. Average distance indicates the average deviation found

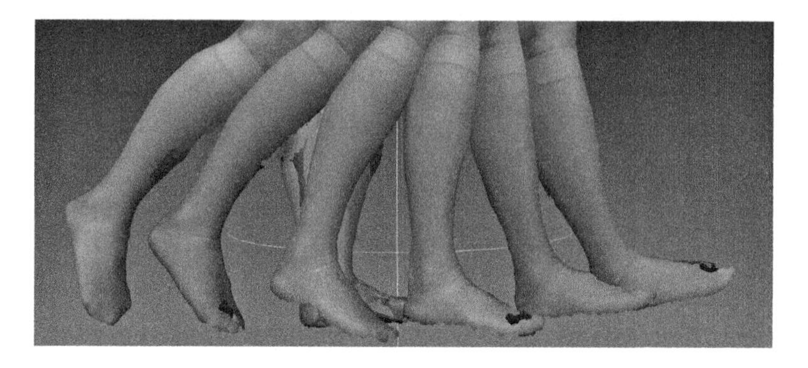

FIGURE 5.2 4D human leg scanning during movement in MeshLab software.

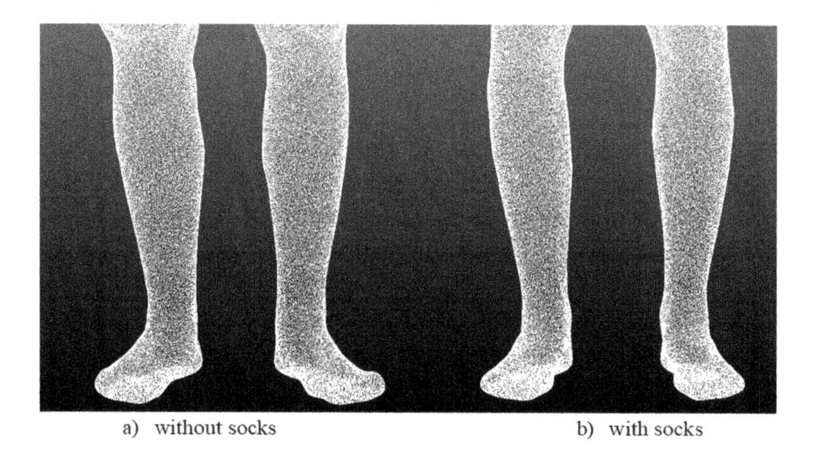

FIGURE 5.3 Point clouds registered results from 4D body scanner of the leg models: (a) without socks and (b) with socks.

from the reference and test leg models. These values can be positive or negative, indicating that the data compared at that area is higher or lower at the surface of the reference model. These values, including the standard deviation, are depicted in Table 5.2.

In order to analyze differences in girth measurements of legs with and without compression socks, section planes at different leg heights are taken. These section planes are further analyzed to calculate girth measurements. Figure 5.6 presents 3D leg models of the participant with section planes. Some of the contour lines visible on the legs are taken at the same heights that correspond to the pressure exerted by the compression socks.

As previously mentioned, every plane taken at different heights on leg models is analyzed to measure girths. In Figure 5.7 are shown view of section planes taken on both legs where 3D models with and without socks are aligned together. Here are

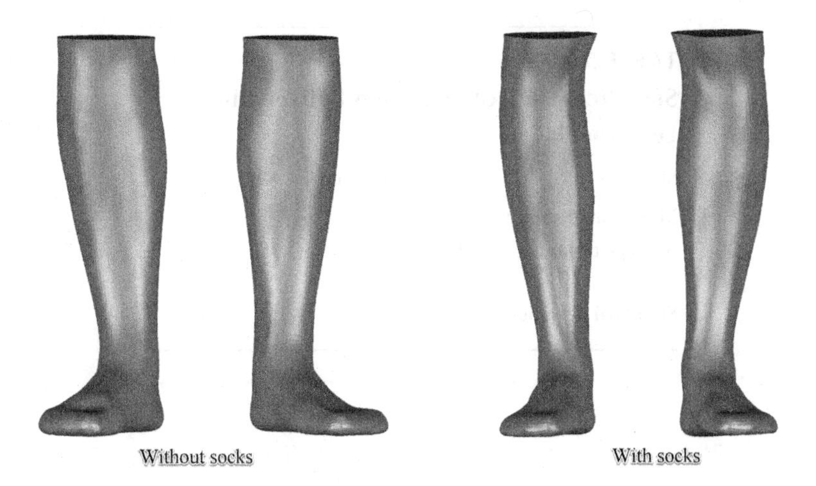

FIGURE 5.4 3D leg models with and without socks.

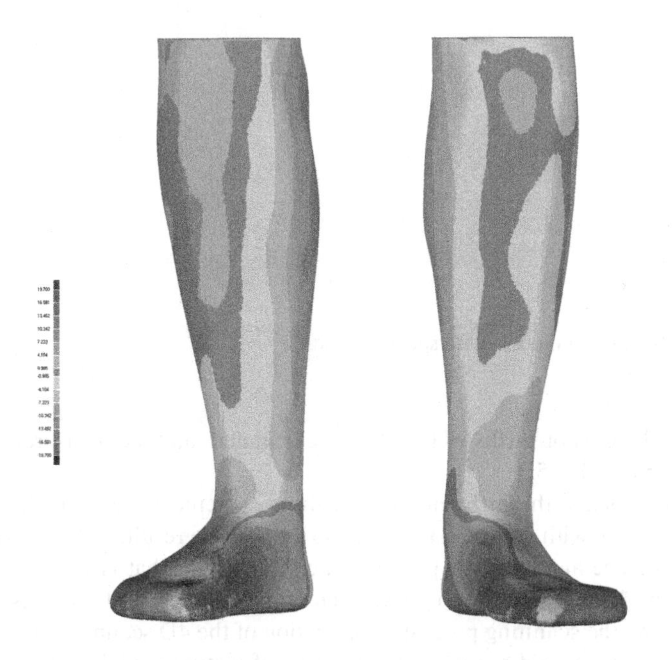

FIGURE 5.5 3D deviation of leg models with and without socks.

visible the differences between both leg models with and without wearing compression socks. In (a) the smallest contour depicts the girth of the right leg with socks and in (c) the largest contour depicts the girth of the right leg without socks. In (b) the smallest contour depicts the girth of the left leg with socks and in (d) the largest contour depicts the girth of the left leg without socks.

TABLE 5.2
Statistical Data of Deviation Between Both Leg Models

Statistics	Positive (mm)		Negative (mm)
Maximum distance	19.7		19.6
Average distance		−1.3	
	2.9		3.6
Standard deviation		4.7	

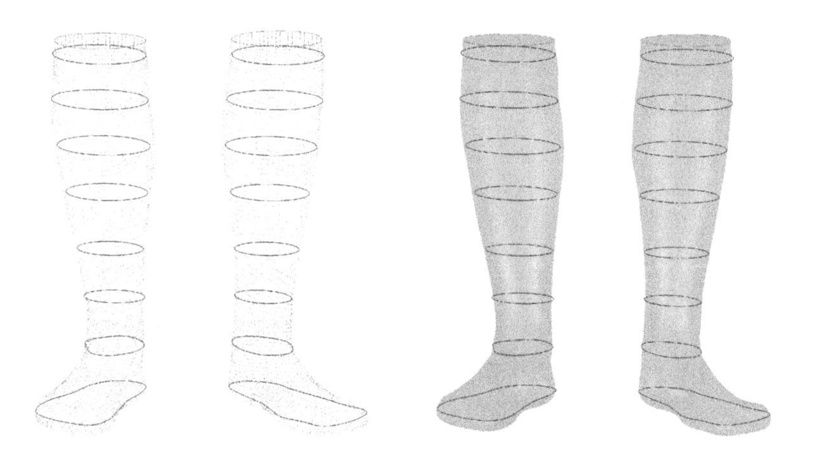

FIGURE 5.6 3D leg models with section planes.

Contour lengths or girths of the leg are calculated, and the measurements taken are depicted in Table 5.3.

From the results, the differences in girth measurements are notable, where the value for the legs with compression socks is lower. As a result, by showing the pressure exerted. The methodology presented here highlights that virtual models generated from the 4D scanning system have been shown to be accurate. This is despite the motion during the scanning process. Application of the 4D scanning process proved to be fast, accurate, and reproducible in terms of virtual models including motion. It offers an accurate 3D mapping and rendering of the human body in movement. Including color and textures, it presents a high level of detail of the body surface. The short time for data capturing is of high importance due to the fact that it eliminates errors from body deformation (Alemany, et al. 2022). Due to the high importance of compression devices, the 4D data of the lower leg improve the design process by ensuring the best fit and at the same time enhancing the comfort of the wearer (Kankariya 2022). Dynamic anthropometry used in ergonomic design improves the

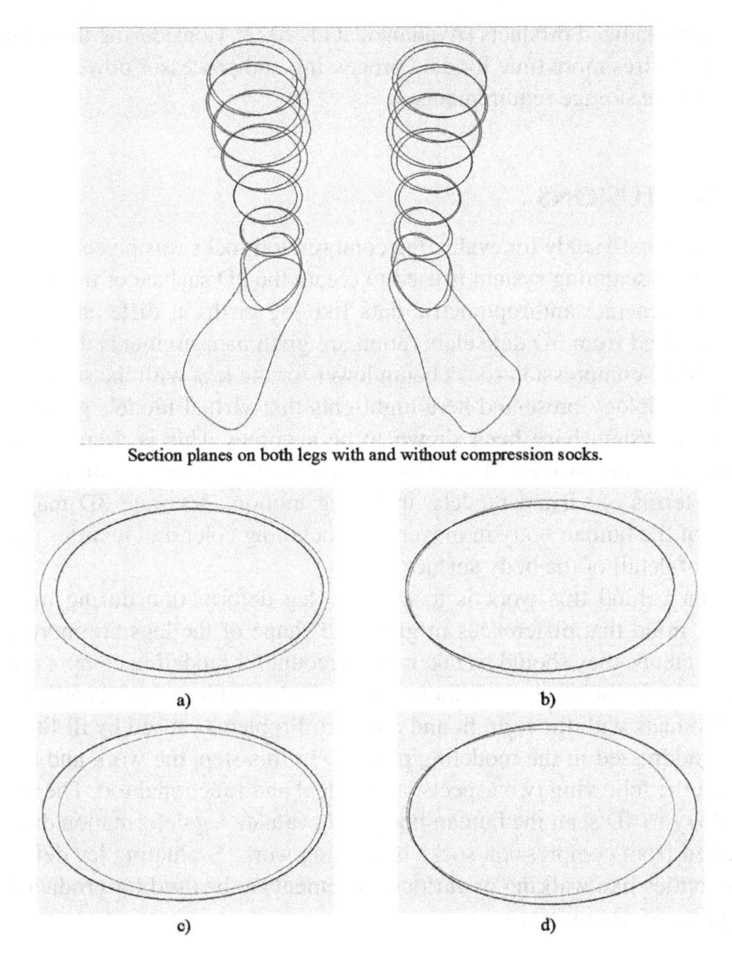

FIGURE 5.7 Section planes of both leg models with and without compression socks. (a) The smallest contour depicts the girth of the right leg with socks, (b) the smallest contour depicts the girth of the left leg with socks, (c) the largest contour depicts the girth of the right leg without socks, and (d) the largest contour depicts the girth of the left leg without socks.

TABLE 5.3
Leg Girth Measurements Without and with Compression Socks

Leg girths without socks (cm)		Leg girths with socks (cm)	
Right leg	Left leg	Right leg	Left leg
35.0	34.5	34.5	34.0
34.2	33.4	32.7	32.9
33.5	33.8	32.5	32.6
29.6	30.0	29.2	29.5
26.4	26.0	26.0	25.5

fitting of personalized products (Avadanei, et al. 2022). Considering that higher data resolution requires more time for data processing and processor power workstations should meet the storage requirements.

5.5 CONCLUSIONS

In this work, a case study for evaluating compression socks was presented. At first, a high-speed 4D scanning system is used to create the 3D surface of the human body and used to generate anthropometric data like leg girths at different heights. The results generated from 3D data elaboration are girth measurements due to the pressure exerted by compression socks being lower for the legs with the socks.

The methodology presented here highlights that virtual models generated from 4D scanning system have been shown to be accurate. This is despite the motion during the scanning process. The 4D scanning process is fast, accurate, and reproducible in terms of virtual models, including motion. Accurate 3D mapping and rendering of the human body in movement, including color and textures, presents a high level of detail of the body surface.

The idea behind this work is to evaluate leg deformation during movements. Bearing in mind that differences in girth and shape of the legs are more prone to body movements, they should be taken into account for modeling compression socks with the right fit. In wide view, footwear products require great attention in order to offer products with the right fit and comfort. Problems caused by ill-fitting shoes should be addressed in the modeling process. In this step, the work and efforts are focused on the following two aspects: aesthetical and functional too. The use of this methodology to 4D scan the human body and evaluate leg deformation due to pressure exerted from compression socks is ongoing work. Evaluating leg deformations due to activities like walking or various movements to be used for product development is part of future work.

5.6 FUTURE TRENDS

Digital technologies have revolutionized the way that many products are developed and offered to customers. From digital design to marketing and buying experience, every process uses these digital tools to improve product development and offer products or services with the right fit and functionality. In the apparel industry, virtual replicas of the human body have been used to gather 3D anthropometric data for a long time. One of the lacks of these data is that body movements are not taken into account in order to predict the right garment fit. 4D body scanning technology proposes a powerful method to capture dynamic anthropometric data that are used for different purposes. A good fit of wearable products can be achieved by taking into account the changes in the human body due to movements and together with various types of materials will improve wearable products used for medicine, space, sports, etc.

ACKNOWLEDGMENTS

The authors would like to thank "German Academic Exchange Service" (DAAD) for the support of Dr. Tatjana Spahiu through the founding program "Research Stays for University Academics and Scientists" 2022 (57588362), at Technische Universität Dresden (TUD), Institute of Textile Machinery and High Performance Material Technology (ITM), Germany. They also would like to thank Dr.-Ing. Kathrin Pietsch, Dipl.-Ing. Ellen Wendt, and Dipl.-Ing. Jessica Boll for their support during this work.

REFERENCES

Alemany, Sandra, et al. 2022. "Data management and processing of 3D body scans." The Textile Institute Book Series. In "*Digital Manufacturing Technology for Sustainable Anthropometric Apparel*", by Norsaadah Zakaria, 97–116. doi: 10.1016/B978-0-12-823969-8.00007-1.

Ashby, Jack, MartinLewis, Roberto SanchisSanchis, Caroline Sunderland, Laura A. Barrett, and John G.Morris. 2021. "Customised pressure profiles of madetomeasure sports compression garments." *Sports Engineering*24 (12). doi: 10.1007/s12283-021-00350-5.

Avadanei, Manuela Lacramioara, SabinaOlaru, IonutDulgheriu, Savin DorinIonesi, Emil ConstantinLoghin, and IrinaIonescu. 2022. "A new approach to dynamic anthropometry for the ergonomic design of a fashionable personalised garment." *Sustainability*. doi: 10.3390/su14137602.

Brophy-Williams, Ned, Matthew W.Driller, Cecilia M.Kitic, and James W.Fell. 2018. "Wearing compression socks during exercise aids subsequent performance." *Journal of Science and Medicine in Sport*22 (1):123–127. doi: 10.1016/j.jsams.2018.06.010.

Brubacher, Kristina, DavidTyler, PhoebeApeagyei, PrabhurajVenkatraman, and Andrew M.Brownridge. 2021. "Evaluation of the accuracy and practicability of predicting compression garment pressure using virtual fit technology." *Clothing and Textile Resarch Journal*. doi: 10.1177/0887302X21999314.

Engel, Florian, ChristianStockinger, AlexanderWoll, and BillySperlich. 2016. "Effects of compression garmentson performance and recoveryin endurance athletes." In *Compression Garments in Sports: Athletic Performance and Recovery*, by Florian Engel and Billy Sperlich. Springer International Publishing. doi: 10.1007/978-3-319-39480-0_2.

Global Compression Wear And Shapewear Market Trends, Strategies, Opportunities For 2022–2031. June 2022. https://www.thebusinessresearchcompany.com/report/compression-wear-and-shapewear-global-market-report.

Gorea, Adriana, and Baytar,Fatma. 2020. "Using 3D body scanning to measure compression variations in a seamless knitted sports bra." *International Journal of Fashion Design, Technology and Education*13 (2):111–119. doi: 10.1080/17543266.2020.1740948.

Ilska, Anita, Krzysztof Kowalski, Magdalena Klonowska, Wojciech Kuzański, Tomasz M. Kowalski, and WitoldSujka. 2017. "Using a 3D body scanner in designing." *Fibres& Textiles in Eastern Europe*25 (5). doi: 10.5604/01.3001.0010.4636.

Kankariya, Nimesh. 2022. "Material, structure, and design of textile-based compression devices for managing chronic edema." *Journal of Industrial Textiles*52 (Jul–Dec 2022):1–35. doi: 10.1177/15280837221118.

Karpiuk, Michal, MaciejMalik, MagdalenaPrzytocka, KatarzynaZcajkovska-Sabat, and WitoldSujka. 2019. "Computer aided manufacturing of compression garments used for rehabilitation of burn and post-operative scars." *Mechanic*1 (2019). doi: 10.17814/mechanik.2019.1.11.

Kisch, Tobias, et al. 2021. "Smart scar care—Industry 4.0 in individualized compression garments: A randomized controlled crossover feasibility study." *Plastic and Reconstructive Surgery – Global Open*15 (9). doi: 10.1097/GOX.0000000000003683.

Kuzmichev, Victor E., Ilia V.Tislenko, and Domenique C.Adolphe. 2018. "Virtual design of knitted compression garments based on body scanning technology and the three-dimensional-to-two-dimensional approach." *Textile Research Journal*89 (12). doi: 10.1177/0040517518792722.

Li, Ziyuan, BennyMalengier, Simona-IleanaVasile, JorisCools, and LievaVan Langenhove. 2019. "From 3D scan to body pressure of compression garments." 19th World Textile Conference on Textiles at the Crossroads, Gent, Belgium.

Parker, Christopher J., SimeonGill, and Stephen G.Hayes. 2017. "3D body scanning has suitable reliability: An anthropometric investigation for garment construction." 8th International Conference and Exhibition on 3D Body Scanning and Processing Technologies, Montreal QC, Canada. doi: 10.15221/17.298.

Salleh, Mohamed Najib Bin, Muhammad Aiman AhmadFozi, and Hendrik BinLamsali. 2017. "The using of 3D handheld scanner to develop a pressure garment model." *Advanced Science Letters*23 (5):4383–4387. doi: 10.1166/asl.2017.8843.

Salleh, Mohamed Najib, Halim MadLazim, Siti NorezamOthman, and Amir FeisalMerican. 2012. "Development of a flexible customized compression garment design system." *International Journal of Advanced Mechatronic Systems*3 (5):202–208. doi: 10.1504/IJAMECHS.2013.057443.

Stephen-Haynes, Jackie, and LouiseToner. 2012. "Outcomes of lower linm 3D scanning and compression hosiery." *Wounds*8 (1):28–34.

Wang, Lijing, MartinFelder, and Jackie Y.Cai. 2011. "Study of properties of medical compression fabrics." *Journal of Fiber Bioengineering & Informatics*4 (1):15–22. doi: 10.3993/jfbi04201102.

Xi, Wang, YangBao, LiQiao, GuoXia, and TaoXiaoming. 2020. "Parametric modeling the human calves for evaluation and design of medical compression stockings." *Computer Methods and Programs in Biomedicine*194 (2020). doi: 10.1016/j.cmpb.2020.105515.

Xiong, Ying, and XiaomingTao. 2018. "Compression garments for medical therapy and sports." *Polymers*10 (6). doi: 10.3390/polym10060663.

Zhi, Gao, LiuTingyu, and Le Conghui. 2015. "The research of Medical Compression Stockings equipment based on 3D scanner." International Conference on Applied Science and Engineering Innovation. doi: 10.2991/asei-15.2015.381.

6 Materials and Design of Pneumatic Compression Devices

Nimesh Kankariya

CONTENTS

6.1 INTRODUCTION

Pneumatic compression devices (PCDs) comprise air-impervious sleeves with an inflatable bladder, a compression pump and connecting tubes. The sleeves come in almost any shape and size to fit the affected physical region (i.e., foot, leg, arm, hand), and the pneumatic compression pumps range in their ability to control pressure, speed and compression method (uniform vs. sequential vs. graduated) (Morris 2008, Morris and Woodcock 2004, Proctor, Greenfield, Wakefield and Zajkowski 2001, Partsch 2008). Detailed classification of PCDs is given in Chapter 1. Materials and design of different components of PCDs are given in the following sections.

6.2 BLADDER

Bladders in PCDs are designed to enclose the specific region of the body, e.g., thigh, calf, foot or combinations of thigh, calf and foot of a human for providing compression treatment to the respective position of the body (Morris 2008, Partsch 2008, Kankariya, Laing and Wilson 2021a). Bladders are made with two sheets (inner and outer bladder layers) of air-impermeable materials (e.g., polyethylene, silicone, polyurethane, nylon-reinforced polyurethane, polyvinyl chloride) (Kankariya, Laing and Wilson 2021b, Kankariya, Wilson and Laing 2021, Kankariya 2022). The inner and outer layers (and consequently the bladder) should not be structurally rigid because

DOI: 10.1201/9781003298526-6

hard, inflexible inner or outer shells may restrict the mobility of the wearer. These two sheets need to be thick enough to withstand thousands of compression cycles without bursting. To protect against puncture and increase durability, inner and outer layers can be laminated or coated. The inner and outer bladder layers are secured together by an adhesive, welding, chemical or mechanical process at the perimeter to form an inflatable chamber, and attached with hook-and-loop addon (e.g., Velcro®), zipper attachments and/or fastening strips. The loop and hook fastening strips can be secured to the inner and outer sheets, respectively (Hasty 1978, Roth, Smith, Brazis and Ramsey 2006).

The bladder, fabricated only with impermeable layers, is uncomfortable for the wearer because of the impermeability of the bladder layers, which traps sensible and insensible water produced by the human body during normal metabolism between the impermeable inner layer and the human skin. Due to this, the patient becomes reluctant to wear the bladder/s or apply the compression using PCDs, compromising their health. A number of researchers attempted to address the issue of sensible and insensible water accumulation on the skin by sealing the inner and outer air impermeable layers to create an air chamber while simultaneously applying uniform pressure to the body (Nicholson and Lipson 1974, Roth et al. 2006, Tordella, Tesluk and Bock 2005). Tordella et al. (2005) developed one such device, in which perforations were punched completely through the top and bottom layers, to facilitate the passage of moisture from the skin of the wearer to the environment thus enabling evaporation of the moisture. However, the arrangement of vents did not provide a satisfactory passage to remove the fluid trapped under the inner layer away from the vent apertures. The evaporation was thus limited to the vent opening and the immediate area near the opening under the air-impermeable layers. A compression device developed by Jacobs et al. (1996) allows for immediate access to the limb of the wearer, but it suffers from the fact that evaporation is still only in specific locations. Sealing the inner soft sheet to an outer rigid sheet and punching the perforations through the sheets were described in Nicholson reference (1974), although the hard and inflexible structure of the outer shell may restrict the freedom of movement of the wearer. According to a Roth et al. (2006), ventilation can be achieved by punching holes of varying sizes through both the outer and inner layers and between adjacent seams. However, perforations at the seam or adjacent to the bladders have the potential to drop pressure, which could lead to plasma pooling.

Three-layer (and four-layer) bladder assemblies have also been fabricated for the lower leg employing various material components, including an extra next-to-skin layer (and an extra outer layer), to facilitate sensible water loss (e.g., Linnane et al. 2014, Roth et al. 2006, Ben-Nun 2013, Brown and Bock 2011, Malhi 2013, Kankariya 2020, Kankariya, Laing, and Wilson 2020). A knit, woven or nonwoven structure composed of textile fibers/filaments (e.g., polyester microfiber, polyester filament, nylon filament, polypropylene filament, wool fiber or blends of wool fiber and nylon filament) was used as an extra next-to-skin layer (and/or for an extra outer cover layer, in the four-layer bladder assemblies). The inner and/or outer layers were either used as a separate layer or fused with a 0.1–0.3 mm thick polyvinyl chloride or polyurethane sheet. Roth et al. (2006), for example, placed a woven nylon fabric, as an innermost layer and in contact with the skin of the wearer during use, on the technical face of the inner layer of the bladder. However, the mode of attachment of woven nylon fabric to the surface of the inner layer was not described. Understanding the technique of attachment is important since laminating the woven material to the sheet might disrupt the arrangement of the fibers, causing them to become discontinuous and reducing moisture transport ability of the material. Kankariya et al. (2021b) developed silicone-based pneumatic bladders for different positions (ankle to above ankle, above ankle to below calf, below calf to calf) on the irregularly shaped lower leg manikin. A combination of drape and flat pattern methods was used to create the patterns for the intermediate layers of pneumatic bladders (Kankariya 2021). The inner intermediate layer was shaped to fit a specified position on the leg manikin. A zip allowance of 20 mm (10 mm on each side of vertical edges) was cut from the basic pattern of the inner intermediate layer. An optimized equal fullness was added to the inner intermediate layer to create the pattern for the outer intermediate layer. The difference in the size of the inner and outer intermediate layers affected the size of the pneumatic bladder and inflation pressure. As the outer intermediate layer was larger than the inner intermediate layer, the inner intermediate layer stretched to coincide with the edges of layers. The edges of the stretched inner intermediate layer and outer intermediate layer were bonded using the primer and adhesive. An open-ended zipper (200 mm, plastic) was bonded to the vertical edges of the bladder. A non-extensible woven plain fabric was cut to the same size as the outer intermediate layer and was bonded to the outer intermediate layer of the bladder to control extension during inflation. Knit tights comprising wool, nylon and elastane were used separately as the innermost next-to-skin layer. In the four-layer bladder assemblies, researchers recommended that perforations be punched in the intermediate bladder layers to allow moisture to pass through (Roth et al. 2006, Kankariya et al. 2021b). The inner layer pulls the moisture from the skin, allowing it to escape through the holes and out the outer layer. The perforations in the layers, on the other hand, may generate a low-pressure zone, lowering the efficacy of the PCDs.

6.2.1 Sealing Methods

Adhesive or chemical bonding seals the edges of the bladder layers by using solvents (to clean the surfaces of the layers before joining) and adhesives (to join the edges of the layers). Wetting between the adhesive and the adherend is required for adhesion to take place. When the surface tension of the liquid adhesive is less than the surface tension of the adherend, wetting, which refers to the spreading out of a liquid on a solid adherend surface, is accomplished. Organic solvents have lower surface tensions than water, which makes it easier for them to wet surfaces. Liquids frequently take on forms that reduce their surface area as a result of forces of attraction (i.e., cohesive forces) between molecules within the polymer. When the contact angle between the adhesive droplet and the adherend surface is less than 60°, good surface wettability is attained; contact angles greater than 90° result in poor surface wettability. Although the surface energy of each adherend is inextricably linked to the polymer compositions of that adherend, the use of the appropriate surface preparation methods (e.g., clean the adherent surfaces to make these free of any impurities that can hinder the forces of attraction; abrade the adherent surfaces to produce mechanical interlocking points and to increase surface area) can compensate for poor wetting. Adhesive or chemical forms attractive forces (e.g., adsorptive force, electrostatic force, diffusive force) between the adhesive and the adherends. The type and strength of attractive force vary with the type of adhesive and adherends. The sort of materials being bonded must be carefully considered while selecting the solvent/adhesive. Since solvents and adhesives are chemical substances, proper handling precautions must be taken when handling them (Kankariya et al. 2021b, Michael 2008).

Radio frequency welding and heat-sealing technique melt and join sealant layers, without the use of solvents or adhesives. Radio frequency welding (also referred to as high-frequency welding) utilizes high-frequency (~ 27.12 MHz) radio waves to produce heat in the materials and ultimately fuse them. A rapidly alternating electric field, set up between two metal welding bars, causes the polarized molecules in the polymer to oscillate and align themselves in relation to the electric field. The heat is generated by the energy that is produced in this process. When this heat is produced with enough energy, the plastic melts and, combined with applied pressure from clamping the welding bars, the layers are sealed together by a free exchange of molecules. The radio frequency energy is then turned off while the plastic sheets are held together by the tooling for a short period of time to cool under pressure. Lowering contamination and improving the capacity to recycle welded components are the advantages of radio frequency welding. However, because it can only be used on materials having polar groups in their chemical structure (the material to be bonded must be able to transform the alternating electric field into heat), radio frequency welding has one major disadvantage (i.e., radio frequency is suitable for polyurethane, polyvinyl chloride; not suitable for rubber, silicone) (Michael 2008, MDDI Online 2022).

6.3 AIR PUMP UNIT

An air pump used in pneumatic compression systems inflates and deflates air-impervious sleeves/bladder that is wrapped around the body. The air pump must reliably generate a desired pressure and maintain it in the bladder and consequently on the body parts for a specified period of time, even during repeated inflation and deflation. The pump system must be able to detect over-inflation of the bladder beyond the specified pressure and decrease pressure by modest deflation. A medical air pump featuring a digital display that shows air pressure, alarm codes (signal the wearer to make appropriate changes in case of over- or under-inflation), and lights to show the current status (e.g., inflation or deflation), a variable air pressure set dial, and user-operated buttons was developed and patented by Orlando et al. (2014a) for use in medical therapy. Furthermore, the adaptability and mobility of PCDs are crucial. A pneumatic compression device with a conventional air pump unit, connected to the bladders, has been developed to provide a substantial volume of air to inflate the bladder and provide the desired compression to the limb (Lina 2003). These devices use a larger compressor which in turn requires a higher power supply and relies on a conventional power connection (with wire – without battery). Wired connection restricts the freedom of participants during treatment (Lina 2003). If the participant connected with a conventional pneumatic compression device needs to use the restroom or move to open a door, the nurse has to release and later reconnect the participant to the device. Thus conventional pneumatic compression devices may be associated with interruption of treatment and are inappropriate for self-use at home, especially by the older population. Other disadvantages of conventional air pumps are related to shape, size and weight. The weight of a conventional pump unit ranges from 2.3 kg to 7 kg (5 pounds to 15 pounds), which makes carrying the system as a portable device difficult and when placed in the vicinity of the participant may impose confinement on the user. These pumps are bulky, which may also create a storage problem in hospitals (Barak, Naveh, and Ehrlich 2002, Fife, Davey, Maus, Guilliod and Mayrovitz 2012). A small, portable mini air pump was developed by Orlando et al. (2014b). It has a rectangular, box-shaped body (which includes a tiny air compressor, air pressure sensor, an electronic circuit board which regulate the functions of the air pump, an AC power connection and a DC battery power supply) with two ports for air supply output, a control panel and a display unit.

6.4 CONNECTING TUBES

When employing PCDs to apply compression, air must be delivered to (inflation condition) and removed from (deflation condition) a bladder of a pneumatic device. Pneumatic devices normally require a direct connection via a tube from the air supply device (air pump unit) to the patient-connected device (bladder). The tube designs might differ in terms of flexibility, compressibility, dimensions (e.g., outer diameter, inner diameter, length) and durability (Orlando and Leonard 2014c). These qualities are determined by the material used to fabricate the tube, with silicone, polyvinyl chloride and polyurethane being the most prevalent options. The tubal flexibility and maneuverability can be enhanced by adding plasticizers or other known additives, but doing so also increases compressibility and, as a result, raises the risk of the tubal lumen occluding. If the patient accidentally compresses the device tube by sitting on, laying on or bending a section of the flexible tubing, the lumen may pinch close and cause the occlusion. When the lumen decreases or collapses, the air transmission slows or stops, which frequently results in a medical emergency (i.e., a prolonged state of increased pressure on the lower leg caused by PCDs in the deflation condition impedes normal blood flow and adds complication to venous disease) or at the very least raises medical concerns (i.e., the device will not apply the predetermined pressure to the body parts in the inflated condition and will not accomplish its purpose). The collapse of the tubal lumen can be prevented by fabricating tubes with a denser, less compressible material – typically metal – coiled within or lined on the inside of their wall. These tubes have the potential to be efficient in limiting occlusion, but their design is substantially more complex, and producing and manufacturing them is more expensive. Orlando and Leonard (2014c) recommended using a connector with a micro bleed hole that connects an air tube to the bladder of the patient. The connector with a microbleed hole safely allows the deflation of a therapy device in the event such device fails due to a blocked air tube or inoperative air pump.

6.5 CONCLUSION AND FUTURE DIRECTIONS

Technological advancements in the material/textile and prototyping industries in recent years have led to the design and development of pneumatic compression systems that integrate complex compression settings (e.g., multi-chamber PCDs which inflate and deflate in wide-ranging patterns and pressures) that have been designed and developed to simulate the muscle tone. Bladders in these pneumatic systems are fabricated by seaming the edges of the air-impervious layers (composed of silicone, polyurethane or polyethylene) and are designed to enclose the specific region of the body, i.e., calf, foot, arm. Because the bladder layers are impermeable to air and water vapor, the bladder is uncomfortable for the wearer, and, hence, the wearer becomes reluctant to wear the bladder/s or PCDs. A number of researchers attempted to address the issue of sensible and insensible water accumulation on the skin by sealing the inner and outer air impermeable layers to create an air chamber while simultaneously applying uniform pressure to the body. Although while using a sweating guarded hot plate, these attempts were the limit examination of the thermophysiological behavior of single-layer material/s and multi-layer assemblies. To further comprehend the thermophysiological and sensory characteristics of the impermeable materials-based PCDs, the study might be expanded to measure the performance properties of assemblies utilizing the thermal manikin technique and human trials.

ACKNOWLEDGMENTS

The author would like to thank Prof. Raechel M Laing and Prof. Cheryl A Wilson for their support.

REFERENCES

Barak, J, Naveh, Y, and Ehrlich, S. 2002. "Portable ambulant pneumatic compression system." US 6494852 B1. Medical Compression System Ltd. Filed Oct 1999, and Issued Dec 2002.

Ben-Nun, A. 2013. "Inflatable compression sleeve." US 0079692 A1. Mego Afek AC Ltd. Filed Nov 2012, and Issued Mar 2013.

Brown, J, and Bock, M G. 2011. "Breathable compression device." US 8029450 B2. Tyco Healthcare Group LP, MA, USA. Filed Apr 2007, and Issued Oct 2011.

Fife, C E, Davey, S, Maus, E A, Guilliod, R, and Mayrovitz, H N. 2012. "A randomized controlled trial comparing two types of pneumatic compression for breast cancer-related lymphedema treatment in the home." *Supportive Care in Cancer*, 20(12), 3279–3286. doi: 10.1007/s00520-012-1455-2.

Hasty, J H. 1978. "Compression sleeve." US 4091804. The Kendall Company, Boston. Filed Dec 1976, and Issued May 1978.

Jacobs, K, Reed, R E, Purdy, W, and Quillen, J B. 1996. "Pressure-normalizing single-chambered static pressure device for supporting and protecting a body extremity." Patent 5489259. Sundance Enterprises Inc. Filed 27/10/1993, and Issued 6/2/1996.

Kankariya, N. 2020. "Characterising a nonwoven component of a textile based compression intervention." AICTE International Conference on "Recent trends in textiles", India.

Kankariya, N. 2021. "Textile and compression of the lower limb." PhD Thesis published at University of Otago, New Zealand.

Kankariya, N. 2022. "Materials, structures, and design of textile-based compression devices for managing chronic edema." *Journal of Industrial Textiles* 52(2022): 1–35. doi: 10.1177/15280837221118844.

Kankariya, N, Laing, R M, and Wilson, C A. 2020. "Challenges in characterising wool knit fabric component of a textile based compression intervention." International Virtuwool Research Conference, New Zealand, AgResearch.

Kankariya, N, Laing, R M, and Wilson, C A. 2021a. "Textile-based compression therapy in managing chronic oedema: Complex interactions." *Phlebology* 36(2), 100–113. doi: 10.1177/0268355520947291.

Kankariya, N, Laing, R M, and Wilson, C A. 2021b. "Prediction of applied pressure on model lower limb exerted by an air pneumatic device." *Medical Engineering & Physics* 97(2021), 77–87. doi:10.1016/j.medengphy.2021.07.007.

Kankariya, N, Wilson, C A, and Laing, R M. 2021. "Thermal and moisture behavior of a multi-layered assembly in a pneumatic compression device." *Textile Research Journal* 92(15–16), 2669–2684. doi: 10.1177/00405175211006942.

Lina, C. 2003. "Pneumatic compression device and methods for use in the medical field." US 0139255 A1. Kinetic Concepts, Inc. Filed Oct 2002, and Issued Jul 2003.

Linnane, P G, Tabron, I S, Fernadez, A, Boström, A L, Hansen, P L, and Mirza, M S. 2014. "Compression device for the limb." US 8636679 B2. Swelling Solution Inc. Filed Oct 2005, and Issued Jan 2014.

Malhi, A S. 2013. "Compression system with vent cooling feature." US 0338552 A1. Tyco Healthcare Group lp. Filed Jun 2012, and Issued Dec 2013.

MDDI online. Accessed on 24 Feb 2022 "Radio frequency welding." [Available from: Radio-Frequency Sealing for Disposable Medical Products (mddionline.com)].

Michael, T. 2008. *Handbooks of plastic joining: A practical guide – II edition*. USA: William Andrew Inc, 2008, pp. 57–63 and pp. 497–505. England: Elsevier.

Morris, R J. 2008. "Intermittent pneumatic compression—Systems and applications." *Journal of Medical Engineering & Technology*, 32(3), 179–188. doi: 10.1080/03091900601015147.

Morris, R J, and Woodcock, J P. 2004. "Evidence based compression: Prevention of stasis and deep vein thrombosis." *Annals of Surgery*, 239(2), 162–171. doi: 10.1097/01.sla.0000109149.77194.6c.

Nicholson, J, and Lipson, C. 1974. "Pressure garment." US 3824992. Clinical Technology Incorporated. Filed Mar 1973, and Issued Jul 1974.

Orlando, M, and Leonard, N. 2014a. "Air pump for use in intermittent pneumatic compression therapy having a digital display." US 9931269. Compression Therapy Concepts Inc. Filed: Mar 2014, Issued: Apr 2018.

Orlando, M, and Leonard N. 2014b. "Compact mini air pump for use in intermittent pneumatic compression therapy." US 20140276290A1. Compression Therapy Concepts Inc. Filed: Mar 2013, Issued: Sep 2014.

Orlando, M, and Leonard N. 2014c. "Micro bleed hole connector for use in intermittent pneumatic compression devices." US 20140276285. Compression Therapy Concepts Inc. Filed: Mar 2013, Issued: Sep 2014.

Partsch, H. 2008. "Intermittent pneumatic compression in immobile patients." *International Wound Journal*, 5(3), 389–397. doi: 10.1111/j.1742-481X.2008.00477.x.

Proctor, M C, Greenfield, L J, Wakefield, T W, and Zajkowski, P J. 2001. "A clinical comparison of pneumatic compression devices: The basis for selection." *Journal of Vascular Surgery*, 34(3), 459–464. doi: 10.1067/mva.2001.117884.

Roth, R B, Smith, B, Brazis, W, and Ramsey, R H. 2006. "Massage device." US 7044924 B1. Midtown Technology, Cleveland, USA. Filed Jun 2000, and Issued May 2006.

Tordella, E, Tesluk, C, and Bock, M. 2005. "Compression apparatus." US 0187503 A1. Brown Rudnick Berlack Israels LLP. Filed Feb 2004, and Issued Aug 2005.

7 Thermal Comfort of Compression Textiles

René M. Rossi

CONTENTS

7.1 INTRODUCTION

As stated by their name, the main purpose of compression garments is to compress human tissue. This implies that the textile is worn in close proximity to the body, without any air layers at the skin-fabric interface. For an optimal performance, the compression garment has to cover the whole or a large part of the limb concerned. On the other hand, the garment has to be sufficiently comfortable to guarantee the acceptance of the wearer. The comfort of a garment is mainly defined by the heat and moisture transfer through the textile (thermal comfort), the sensory or tactile comfort, as well as the garment fit. In the case of compression garments, the fit can be mostly characterized by the pressure applied by the garment to the skin. The tight fit as well as the high coverage factor can be a challenge in terms of thermal comfort, especially in summer, as the compressing structure requires a minimal thickness and density, which may hinder the evacuation of the heat and moisture produced by the body. The heat production of the human body during sports activities can be very high (above 1000 W), and, therefore, the textiles have to be designed to allow

DOI: 10.1201/9781003298526-7

sufficient heat and moisture transfer to avoid overheating of the body. On the other hand, compression textiles for medical applications will usually be used for more sedentary activities with less heat production. However, people using such garments (e.g. elderly people) may suffer from thermoregulatory impairments, and therefore, the balance between heat production and heat loss in the body has to be constantly maintained. To obtain good thermophysiological comfort, the textile has to fulfill several criteria: it has to be permeable to the transfer of dry heat and water vapor, and, if the body sweats in liquid form, it has to allow fast evaporation to cool the body efficiently. This chapter gives an overview of the thermal interactions between the body, the clothing, and the environment and describes the properties of compression garments necessary to ensure good thermophysiological comfort for the wearer.

7.2 HUMAN THERMOREGULATION AND HEAT EXCHANGE MECHANISMS

The human body, similarly to all homeothermic animals, has to maintain its temperature closely around 37°C. This means that body heat production should optimally be equal to body heat loss. The body constantly produces heat, mainly in the organs and the muscles. This heat production can vary from about 80 W at rest up to over 1000 W during strenuous sports activities like a marathon. The body temperature is regulated by the hypothalamus that collects the temperature signals from different sensors in the body core and on the skin, and compares it to a so-called set-point temperature. If the overall body temperature is below this set-point temperature, the blood vessels in the periphery (skin) contract (vasoconstriction), which limits the blood flow to the skin and thus prevents excessive heat loss. Furthermore, the body has mechanisms to increase metabolic heat production, mainly through muscles shivering. On the other hand, if the body temperature is above the set-point temperature, the skin blood vessels dilate (vasodilatation), which increases the mean skin temperature and thus the dry heat transfer to the environment. If the body temperature increases beyond a certain threshold, the sweat glands produce liquid sweat, which cools the body when it evaporates. Furthermore, the body constantly loses water vapor through the diffusion of the water molecules through the skin (transepidermal water loss – TEWL). This TEWL is around 20 g/h in normal thermal conditions (20°C).

The thermal environment around the body can be characterized by four parameters: the air temperature, the radiant temperature, the relative humidity in the air, and the air velocity. The heat exchange between the body and the environment can occur by different means: dry heat exchange, evaporative heat loss, as well as heat loss through breathing. Dry heat can be transferred by three different means:

1. Conduction is defined as the heat transfer in an object from the region with a high temperature to the region with a lower temperature, or between two objects of different temperatures in physical contact. The heat transfer is dependent on a material property called thermal conductivity.

2. Convection describes the motion of a fluid due to a density difference. In the case of convective heat transfer, this density difference in the fluid is due to a temperature difference. Convection is divided into two different types: natural convection occurs when no external source produces the fluid motion, i.e. the fluid movement is only produced by the temperature-dependent density difference. On the other hand, forced convection occurs under the action of an external force. Convective heat transfer between the human body and the environment can thus either occur in air or in water.
3. Radiation is produced by any object with a temperature above absolute zero. Electromagnetic waves are emitted by the object and the radiant heat transfer occurs across any transparent medium. The emission of electromagnetic waves by a material is dependent on its emissivity. The emissivity of the human skin is 0.98, while it can vary between about 0.7 and 0.95 for common textiles.

When the thermal environment reaches temperatures nearly equal to the skin temperature, the dry heat loss is reduced to zero. In the case of environmental temperatures higher than skin temperature, the dry heat transfer turns into a heat gain for the body. For such scenarios, the human body can only lose heat by sweat evaporation. Therefore, the garments used in such conditions have to promote this evaporation as much as possible. This means that the liquid sweat should evaporate as near as possible from the skin, as the skin cooling efficiency is very dependent on the distance from the skin (Havenith et al. 2013, Wang et al. 2014). However, the sweat should not accumulate near the skin as such a liquid accumulation can alter the properties of the skin (softening and thickening of the upper skin layers) (Dąbrowska et al. 2016). Furthermore, moisture at the interface alters the frictional properties between the textile and the skin that may lead to a reduction of sensorial comfort or even friction-related injuries like blisters (Derler, Rossi and Rotaru 2015).

7.3 MEASUREMENT METHODS TO ASSESS THE THERMAL AND MOISTURE TRANSFER THROUGH TEXTILES

7.3.1 Fabric Tests

In order to optimize the thermal comfort of textiles, materials parameters like thermal conductivity or heat capacity can be assessed. One of the most common methods to measure the thermal resistance as well as the water vapor resistance of a textile is the sweating guarded hotplate according to the International Standard ISO 11092 (Textiles – Physiological effects – Measurement of thermal and water-vapor resistance under steady-state conditions (sweating guarded-hotplate test)). This method consists of a porous metal plate heated to 35°C located in a climatic chamber. The textile samples are placed directly onto the plate without an air gap and a fan blows air with a constant velocity (1 m/s) tangentially over the sample to ensure reproducible conditions. The heating power to maintain the metal plate at a constant temperature is measured continuously. For the measurement of the thermal resistance

R_{ct}, the temperature in the climatic chamber is regulated at 20°C, with a relative humidity of 65%. The heating power is recorded once it reaches steady-state and the thermal resistance is calculated by multiplying the temperature difference between the plate and the climatic chamber (15°C) by the measuring area of the plate, divided by the supplied steady-state heating power and is given in the SI unit m^2K/W. For water vapor resistance, water is supplied to the porous metal plate from underneath. The plate is covered by a waterproof and water-vapor permeable cellophane foil to prevent the wetting of the sample. The climate in the chamber is adjusted at 35°C and 40% relative humidity to obtain isothermal conditions and thus to avoid any problems of moisture condensation in the sample. The water evaporates in the plate and flows through the cellophane and the sample. Due to the evaporative cooling of the water, the plate has to be constantly heated to maintain 35°C. The water vapor resistance (in m^2Pa/W) can be calculated by multiplying the difference of water vapor partial pressure between the chamber and the plate by the measuring area of the plate divided by the supplied steady-state heating power.

Gravimetric methods to determine the water vapor transfer through a textile are also commonly used. Such a method is described in the American Standard ASTM E96 (Standard Test Methods for Water Vapor Transmission of Materials). The fabric sample is placed over a cup filled with either water or a desiccant. The cup is then placed on a scale in a climatic chamber with a defined temperature and relative humidity. The weight change of the cup (decrease when filled with water and increase when containing the desiccant) is constantly measured and the water vapor transmission rate (in g/m^2h) can be calculated.

When the body starts producing sweat as a reaction to a too-high body temperature, this liquid has to evaporate as fast as possible to cool down the skin and the body core. As the body does not sweat homogeneously (Smith and Havenith 2012), a near-to-body textile has to be able to adsorb the sweat and distribute it over a large surface. This wicking ability of a fabric can be measured with different standard methods. One of the most common methods is the moisture management tester (MMT) described in AATCC 195 (Liquid Moisture Management Properties of Textile Fabrics). The fabric is placed horizontally between two electrical sensors with seven concentric pins. A defined amount of liquid is then poured on the fabric and the spreading is recorded through a change in electrical conductivity. Thus, the specific spreading at the top and/or the bottom surface can be measured as well as the wicking of the liquid from the top surface to the bottom surface.

7.3.2 Body and Body Part Models (Manikins)

Fabrics tests are important to determine materials' parameters. However, they do only partly reproduce the heat and mass transfer in a garment, as they usually do not take into consideration the interactions between these two mechanisms. The evaporation of moisture necessitates thermal energy while condensation leads to heat release. Therefore, heat and mass transfer have to be assessed simultaneously to realistically determine their influence on the thermal comfort of the wearer. Furthermore, additional parameters like geometrical factors or the design of the clothing (e.g. stretching of the fabrics during wear) have to be considered as well. For this reason, several test methods (so-called manikins) have been developed to reproduce the geometry of the human body or single body parts like the head (Bogerd et al. 2015, Psikuta et al. 2017) or the foot (Bogerd, Brühwiler and Rossi 2012). These manikins are usually divided into different segments representing the body parts that can be separately heated. State-of-the-art manikins have sweating systems, either integrated into the segments (sweating nozzles) or by means of a wet fabric put onto the manikin. The surface of the manikin is usually regulated to a defined temperature and thus is not directly simulating the human thermoregulation mechanisms (vasoconstriction or shivering) (Bogerd et al. 2010). However, different numerical models of human thermoregulation with high spatial and temporal resolutions have been recently developed (Psikuta et al. 2017), which have been coupled with the thermal manikins. Such adaptive manikins are called "human simulators". Thus, the surface temperature or the heating power of the manikin is no longer set to a constant value but adjusted according to calculated values from the numerical model. Two methods can be used to control this adaptive manikin:

- The overall surface heat flux measured by the manikin is used as input data for the model, and the calculated surface temperature and sweat rate are used as control parameters of the manikin.
- The heat flux and the sweat rate are the controlled parameters on the manikin, and the measured manikin surface temperature is the feedback parameter for the numerical model.

Human simulators are convenient tools to realistically assess the heat and moisture transfer properties of garments. However, as the numerical thermophysiological models are based on empirical data from human trials, their validity is limited to specific use scenarios. Therefore, these simulators can usually not be used to assess complex (heterogeneous) heat and mass transfer mechanisms, such as rapidly changing thermal environments or garments with strongly varying insulations in the different body parts.

7.4 HEAT AND MOISTURE TRANSFER PROPERTIES OF COMPRESSION FABRICS AND GARMENTS

Compression textiles are usually made of a materials blend with heterogeneous yarn materials (Liu et al. 2017). The yarns usually consist of a yarn wrapped around a core yarn made of an elastic fiber (Spandex). The wrapped yarns can be made of different fiber types like cotton or polyamide that largely influence the moisture transport and absorption properties of the garment. Most compression socks and stockings are made with a circular knitting machine for seamless garments. The knitting parameters have a major effect on the thermal comfort of the wearer as differences in fiber types, as well as yarn and fabric structures affect the fabric thickness and density. The thickness of the fabric is inversely proportional to the thermal conductivity and the density correlates with the size of the pores in the fabric and thus with the air permeability (and usually also with the moisture permeability). In order to reach the level of pressure required, the compression garment has to have a minimal thickness and density, which hinders body heat and moisture evacuation and therefore reduces thermal comfort. Furthermore, compression garments have to cover whole body sites to reach the necessary pressure on the human tissue and underlying vascular system, which can also lead to a reduction of body heat and moisture transfer. Thus, the requirements to reach the desired compression effect are contradictory to those for high thermal comfort and a balance between these conflicting properties has to be found.

7.4.1 Fabrics Properties

It could be shown that the thermal resistance of a knit is strongly correlated to its thickness, mass per unit area, cover factor, and porosity but not to the fiber thermal conductivity (Čubrić et al. 2012). Thus, the amount of air entrapped in the knits is more important for the thermal insulation of the textile than the material. The effect of the yarn type on the performance of diabetic socks was recently analyzed by Cüreklibatır Encan and Marmaralı (2022). Apart from thermal properties (thermal conductivity, air, and water vapor permeability), they also analyzed the coefficient of friction and the recovery after compression of cotton, polyamide, lyocell, polyester, acrylic, viscose, and their blends. They concluded that the polyester yarn showed the best compromise with high air and water vapor permeability with a low coefficient of friction and a high recovery rate after compression.

Thermal comfort, especially during sports activities, is strongly related to water vapor resistance and moisture accumulation within the clothing (Fan and Tsang 2008). The fabric liquid absorption capacity and the drying rate could be related to the fabric density (Kaplan and Akgünoğlu 2021). For activities with high metabolic heat production or at high environmental temperatures, the body produces liquid sweat and the wicking of this liquid in the textile structure has to be optimized. Wicking is defined as the spontaneous liquid imbibition due to capillary forces in the presence of a hydrophilic fiber system, pulling the liquid into the pore space (Parada et al. 2017). It primarily occurs within yarn's pore spaces formed between hydrophilic fibers and is influenced by yarn parameters such as diameter, twisting level, and fiber density. The transport of liquid from yarn to yarn essentially occurs at the contact points between the yarns and not through the void space between yarns (Kim, Michielsen and DenHartog 2021). It could be recently shown that the yarn-to-yarn liquid transition is characterized by a slow water advance at the yarn contact, but that the time for the transition cannot be predicted precisely due to the irregular fiber arrangement in yarns (Fischer et al. 2022).

Manshahia and Das analyzed the thermal comfort properties of different knitted active sportswear and concluded that comfort is significantly affected by the structural parameters (Manshahia and Das 2013). The analyzed sportswear were nine different tennis and soccer t-shirts made of multi-filament polyester-knitted fabrics with different filament cross sections (elliptical, hexagonal, triangular, and circular) and knit structures (interlock, plaited, two-layer). They could show that the interlock fabrics had a higher water vapor permeability in comparison to the other structures. The filament cross-section influenced the specific surface area of the filaments and thus the capillary pressure in the intra-yarn capillaries. In consequence, the filaments with a higher shape factor led to a better wicking effect.

Birrfelder et al. (2013) analyzed the intra- and inter-yarn transport of liquid water of different knit structures when the textiles were subjected to different mechanical pressures and showed that the amount of liquid supplied to the textile influenced the water transport. In a supplied-limited flow, the water could be adsorbed by the capillary structure and the transport was driven by capillary forces. However, when excessive water was present (supply-driven flow), the liquid was being pressed into the fabric. They could show that different waterfronts were present in the knit structures that could be attributed to the intra-yarn and inter-yarn wicking. The in-plane wicking rate was the highest for interlock-type knits compared to rib and eyelet structures. In an earlier study, the same group showed that applying mechanical pressure generally increased the in-plane water transport (Rossi et al. 2011). However, if the increased pressure leads to a strong reduction of the void spaces, liquid movement can be obstructed. Kumar et al. (2015) showed that the liquid transmission in nonwoven padding bandages was reduced with increasing pressure, causing a higher accumulation of fluid in the textile structure. Samples with a lower density showed a higher rate of liquid spreading.

7.4.2 Heat and Moisture Transfer in Garments

The heat and moisture transfer properties in garments can greatly vary from the properties measured on the fabric level due to garment design features like air layers or garment openings. During wear, walking movements largely change the convective heat exchange (ventilation), and, therefore, the effects of air permeability on the effective thermal insulation of the garment have to be analyzed in realistic wear scenarios (Morrissey and Rossi 2014). Compression garments are tight-fitting, and, therefore, the materials are subjected to a certain degree of stretching that changes the porosity and thus the form of the capillaries between fibers and yarns, which influences the air and moisture permeability as well as the wicking properties of the garment. Priyalatha and Raja (2017) investigated the wicking behavior of several cotton knits under different dynamic deformation states and showed that the wicking significantly increases compared to the static, non-deformed state. Wang et al. (2021) studied the strain-dependent vertical and horizontal wicking behavior of different cotton/lycra elastic woven fabrics and showed that both the vertical wicking height and the horizontal wicking area and wicking rates were increased for higher strains. The same group could also show a dependency of the liquid spreading area with tensile strain for warp stretch woven fabrics made of nylon/spandex/cotton (Wang et al. 2022).

In several studies, the coupled heat and moisture transport properties of socks have been analyzed with foot manikins. This measurement method allows the assessment of design features of footwear like ventilation openings and the quantification of heat loss, water vapor permeability as well as moisture absorption in the socks for different foot regions (Bogerd, Brühwiler and Rossi 2012, Kuklane 2004, Castellani et al. 2014). However, the comparison of manikin analyses with human subject trials showed that the subjective evaluations did not seem to have the same sensitivity and discriminative power between different garments (West et al. 2020).

The effect of heat and moisture transport properties of a garment on skin health (i.e. transepidermal water loss and hydration of the stratum corneum) was analyzed in several studies. Laing et al. (2015) showed that wool socks were associated with improved skin health compared with acrylic or cotton socks, although the water vapor transfer of the fabrics was not different. In a more recent study, the effect of single jersey socks with different fiber types (cotton, wool, polyester, coolmax) on the perception of thermal comfort was evaluated and compared with the effect of not wearing a sock (West, Havenith and Hodder 2021). The foot skin temperature and shoe microclimate were measured at seven sites on the right foot and foot skin hydration on nine sites. The study showed similar thermophysiological and perceptual responses for all sock types. However, foot wetness, stickiness, and discomfort were perceived as higher when no sock was worn. Other studies also reported no significant differences in thermophysiological parameters (skin temperature and sweat rate) and thermal perception for different sock types (cotton/nylon vs. cotton/coolmax/polypropylene/nylon (Purvis and Tunstall 2004); cotton vs. polyester/cotton/olefin/spandex vs. polyester/spandex (Barkley et al. 2011); cotton vs. olefin (Van Roekel, Poss and Senchina 2014)). However, results of human subject trials might vary depending on the experimental conditions (e.g. duration of the exercise, thermal properties of the shoes). This was demonstrated in the studies by Bogerd, Niedermann, Brühwiler, Rossi (2012) who found greater skin hydration for a polypropylene/elasthane sock compared to a blend (wool/polypropylene/polyamide) during a long-duration exercise in boots with low water vapor permeability in the field, but no significant difference in an earlier study with the same sock types in the laboratory (Bogerd et al. 2011). Differences in foot skin temperatures were also found in a recent study with 30 male athletes who ran a half-marathon wearing either polyester socks or socks with polyester and bioceramic fibers (Escamilla-Martínez et al. 2022). The socks with bioceramic fibers led to a lower skin temperature (1.1–1.3°C) than the polyester socks.

7.4.3 Optimization of Heat and Moisture Transfer in Garments

In recent years, research in thermophysiology has largely concentrated on the development of new materials to improve the thermal and moisture management of garments. Different new concepts have been proposed to optimize either the dry heat transfer or maximizing the efficiency of evaporative cooling. These developments have not been specifically made for compressive garments, but could potentially be used for this purpose. This section shortly reviews these strategies, and more information on this topic can be found in specific reviews (Lou, Chen and Fan 2021, Farooq and Zhang 2021).

The thermal insulation of tight-fitting compression garments is often too high. To increase the heat loss from the body to the environment, materials with high thermal conductivity can be integrated into the garment. For example, cotton or polyester fabrics were coated with multiwalled carbon nanotubes or with nano graphene powders to increase the thermal conductivity (Abbas et al. 2013, Manasoglu et al. 2019). Phase change materials (PCM) have also been proposed in sportswear to regulate body heat loss (Nusser and Senner 2010). These materials change their aggregation state from solid to liquid and can thus store or release heat during the phase transition process.

In order to increase the radiant heat loss, several so-called infrared transparent visible-opaque (ITVO) fabrics have been developed. Normal clothing has an emissivity of around 0.7–0.95, which is near the emissivity of the skin in the mid-infrared range of around 7–14 μm (0.98) and, therefore, has a very low light transmission in the infrared region. By using polymers with high infrared transmittance (such as polyethylene), garments with a high transmittance (0.97) can be obtained (Lou, Chen and Fan 2021).

To guarantee efficient evaporative cooling, the in-plane and transplanar wicking effects of the garments have to be optimized. This can usually be obtained with hydrophilic treatments of the fibers that improve water transport. Different cooling garments have been proposed for medical and sports applications. Meyer-Heim et al. developed a lightweight cooling garment consisting of two semipermeable membranes and a wicking fabric between the two membranes (Meyer-Heim et al. 2007). By wetting the wicking fabric with water, human sweating can be mimicked and, thus, the evaporative cooling effect could be maximized. Other types of air-cooled or liquid-cooling garments have been proposed, mainly to be worn underneath protective clothing (Mokhtari Yazdi and Sheikhzadeh 2014), but are probably less appropriate for tight-fitting garments.

7.5 FUTURE TRENDS: OPTIMIZATION OF THE THERMAL COMFORT AND THE WEARABILITY

Future research will continue to optimize the balance between the compression function, the thermal comfort, and the wearability of compressive textiles. In terms of wearability, the main problem is the easy donning and doffing of such textiles. New technologies in smart textiles can provide interesting options to optimize this balance between stabilizing properties, thermal comfort, and ergonomics.

7.5.1 Active Textiles

Inflatable or liquid-filled structures are a promising alternative to the traditional compressive textiles containing elastic fibers like Spandex (Chung et al. 2018; Kankariya, Laing and Wilson 2021, Zhu et al. 2022). These active compression garments usually contain several airtight bladders (or liquid-filled tubes) and a soft pneumatic actuator and the pressure applied to the limb can be varied, for example for intermittent compression therapy (Zhao et al. 2020). However, in such systems, the bladders are usually not only airtight but also impermeable to water vapor, which increases the risk of moisture accumulation at the skin-garment interface and limits the evaporative heat loss. Recently, a novel laser welding technology was developed that enables the bonding of semipermeable membranes to obtain breathable inflatable structures (Fromme et al. 2020). This new technology may be used in the future to produce inflatable bladders and can thus contribute to the improvement of both the sensorial and thermal comfort of compression garments.

Apart from inflatable systems that require a non-textile, rather bulky inflation system, compression garments containing shape-changing materials that can produce a controllable compression have been recently proposed (Duvall et al. 2017, Narayana et al. 2017). Textiles containing shape memory alloys (SMA) should facilitate the donning at room temperature in a low-stiffness state and provide compression when the SMA makes a transition to a high-stiffness state at the higher skin temperature (Granberry et al. 2022). Such systems are thinner and less dense than classical compression textiles and, therefore, could potentially offer better thermal comfort to the wearer.

7.5.2 Sensing Textiles

Smart textiles used for the monitoring of vital signs are finding applications in the field of compression textiles. It was recently shown that monitoring the foot temperature of diabetic patients can be a predictor of diabetic foot syndrome and a prototype for such a smart sock with integrated thermistors was proposed (Torreblanca González et al. 2021). However, such smart garments have to be non-intrusive to be accepted by the patients, as wearable electronics only mounted on the textiles may be uncomfortable (Drăgulinescu et al. 2020). A promising approach for a full embedding of the sensor is the use of optical fibers that can be directly integrated into the sock structure (Guignier et al. 2019). Different temperature sensors based on optical fibers and Fiber Bragg Gratings (FBG) using a selective photo-induced modulation of the core refractive index have been developed (Bilro et al. 2012, Roriz et al. 2020). First studies with patients suffering from diabetic peripheral neuropathy have shown the validity of such smart socks for the assessment of plantar temperature in a clinical setting (Najafi et al. 2017), resulting in significantly different temperatures in comparison to a control group (Reyzelman et al. 2022, Scholten et al. 2022).

7.6 CONCLUSION

The thermal comfort of compression garments is usually not optimal due to the use of rather thick and dense fabrics to obtain the necessary compression during use. Furthermore, such garments cover large areas of the concerned limbs. In such situations, the evacuation of the heat and moisture produced by the body is usually limited leading to a reduction of thermal comfort during wear. The use of novel materials allowing a better radiant heat transfer or coatings that increase the thermal conductivity through the textiles could help improve the heat exchange with the environment. Apart from the optimization of the dry heat transport, the distribution and fast evaporation of the sweat produced by the body in the compressive textiles can support heat evacuation and avoid the storage of moisture at the skin-garment interface. New materials with adaptable shape are promising developments that not only facilitate the donning of compression garments but can also improve thermal comfort with the use of lighter and thinner textiles.

REFERENCES

Abbas, Amir, Yan Zhao, Xungai Wang, and Tong Lin. 2013. "Cooling effect of MWCNT-containing composite coatings on cotton fabrics." *Journal of The Textile Institute* 104 (8):798–807.

Barkley, Rachel M, Mike R Bumgarner, Erin M Poss, and David S Senchina. 2011. "Physiological versus perceived foot temperature, and perceived comfort, during treadmill running in shoes and socks of various constructions." *American Journal of Undergraduate Research* 10 (3):7–14.

Bilro, Lúcia, Nélia Alberto, João L Pinto, and Rogério Nogueira. 2012. "Optical sensors based on plastic fibers." *Sensors* 12 (9):12184–12207.

Birrfelder, Pascale, Marko Dorrestijn, Christian Roth, and René M Rossi. 2013. "Effect of fiber count and knit structure on intra-and inter-yarn transport of liquid water." *Textile Research Journal* 83 (14):1477–1488.

Bogerd, Cornelis P, Jean-Marie Aerts, Simon Annaheim, Peter Bröde, Guido De Bruyne, Andreas D Flouris, Kalev Kuklane, Tiago Sotto Mayor, and René M Rossi. 2015. "A review on ergonomics of headgear: Thermal effects." *International Journal of Industrial Ergonomics* 45:1–12.

Bogerd, Cornelis P, Paul A Brühwiler, and René M Rossi. 2012. "Heat loss and moisture retention variations of boot membranes and sock fabrics: A foot manikin study." *International Journal of Industrial Ergonomics* 42 (2):212–218.

Bogerd, Cornelis P, Reto Niedermann, Paul A Brühwiler, and René M Rossi. 2012. "The effect of two sock fabrics on perception and physiological parameters associated with blister incidence: A field study." *Annals of Occupational Hygiene* 56 (4):481–488.

Bogerd, Cornelis Peter, Ivo Rechsteiner, Benno Wüst, René M Rossi, and Paul A Brühwiler. 2011. "The effect of two sock fabrics on physiological parameters associated with blister incidence: A laboratory study." *Annals of Occupational Hygiene* 55 (5):510–518.

Bogerd, N, A Psikuta, HAM Daanen, and RM Rossi. 2010. "How to measure thermal effects of personal cooling systems: Human, thermal manikin and human simulator study." *Physiological Measurement* 31 (9):1161.

Castellani, John W, Robert Demes, Thomas L Endrusick, Samuel N Cheuvront, and Scott J Montain. 2014. "Heat removal using microclimate foot cooling: A thermal foot manikin study." *Aviation, Space, and Environmental Medicine* 85 (4):445–448.

Chung, Jinwon, Roman Heimgartner, Ciaran T O'Neill, Nathan S Phipps, and Conor J Walsh. 2018. "Exoboot, a soft inflatable robotic boot to assist ankle during walking: Design, characterization and preliminary tests." 2018 7th IEEE International Conference on Biomedical Robotics and Biomechatronics (Biorob).

Čubrić, I Salopek, Zenun Skenderi, Alka Mihelić-Bogdanić, and Mladen Andrassy. 2012. "Experimental study of thermal resistance of knitted fabrics." *Experimental Thermal and Fluid Science* 38:223–228.

Cüreklibatır Encan, B, and A Marmaralı. 2022. "Effect of yarn type on performance of diabetic socks." *Indian Journal of Fibre & Textile Research (IJFTR)* 47 (3):290–295.

Dąbrowska, Agnieszka K, Christian Adlhart, Fabrizio Spano, Gelu-Marius Rotaru, Siegfried Derler, Lina Zhai, Nicholas D Spencer, and René M Rossi. 2016. "In vivo confirmation of hydration-induced changes in human-skin thickness, roughness and interaction with the environment." *Biointerphases* 11 (3):031015.

Derler, S, RM Rossi, and GM Rotaru. 2015. "Understanding the variation of friction coefficients of human skin as a function of skin hydration and interfacial water films." *Proceedings of the Institution of Mechanical Engineers, Part J: Journal of Engineering Tribology* 229 (3):285–293.

Drăgulinescu, Andrei, Ana-Maria Drăgulinescu, Gabriela Zincă, Doina Bucur, Valentin Feieş, and Dumitru-Marius Neagu. 2020. "Smart socks and in-shoe systems: State-of-the-art for two popular technologies for foot motion analysis, sports, and medical applications." *Sensors* 20 (15):4316.

Duvall, J., R. Granberry, L. E. Dunne, B. Holschuh, C. Johnson, K. Kelly, B. Johnson, and M. Joyner. 2017. The design and development of active compression garments for orthostatic intolerance. In *Frontiers in Biomedical Devices, BIOMED - 2017 Design of Medical Devices Conference, DMD 2017.*

Escamilla-Martínez, Elena, Beatriz Gómez-Martín, Raquel Sánchez-Rodríguez, Lourdes M Fernández-Seguín, Pedro Pérez-Soriano, and Alfonso Martínez-Nova. 2022. "Running thermoregulation effects using bioceramics versus polyester fibres socks." *Journal of Industrial Textiles* 51 (8):1236–1249.

Fan, Jintu, and Humble WK Tsang. 2008. "Effect of clothing thermal properties on the thermal comfort sensation during active sports." *Textile Research Journal* 78 (2):111–118.

Farooq, Abdul Samad, and Peng Zhang. 2021. "Fundamentals, materials and strategies for personal thermal management by next-generation textiles." *Composites Part A: Applied Science and Manufacturing* 142:106249.

Fischer, Robert, Christian M Schlepütz, René M Rossi, Dominique Derome, and Jan Carmeliet. 2022. "Wicking through complex interfaces at interlacing yarns." *Journal of Colloid and Interface Science* 626:416–425.

Fromme, Nicolas Philip, Martin Camenzind, Robert Riener, and René M Rossi. 2020. "Design of a lightweight passive orthosis for tremor suppression." *Journal of NeuroEngineering and Rehabilitation* 17 (1):1–15.

Granberry, Rachael, Megan Clarke, Robert Pettys-Baker, Heidi Woelfle, Crystal Compton, Amy Ross, Kirstyn Johnson, Santo Padula, Surbhi Shah, and Julianna Abel. 2022. "Dynamic, tunable, and conformal wearable compression using active textiles." *Advanced Materials Technologies* 7(12):2200467.

Guignier, Claire, Brigitte Camillieri, Michel Schmid, René M Rossi, and Marie-Ange Bueno. 2019. "E-knitted textile with polymer optical fibers for friction and pressure monitoring in socks." *Sensors* 19 (13):3011.

Havenith, George, Peter Bröde, Emiel den Hartog, Kalev Kuklane, Ingvar Holmer, Rene M Rossi, Mark Richards, Brian Farnworth, and Xiaoxin Wang. 2013. "Evaporative cooling: Effective latent heat of evaporation in relation to evaporation distance from the skin." *Journal of Applied Physiology* 114 (6):778–785.

Kankariya, Nimesh, Raechel M Laing, and Cheryl A Wilson. 2021. "Prediction of applied pressure on model lower limb exerted by an air pneumatic device." *Medical Engineering & Physics* 97:77–87.

Kaplan, Sibel, and Betül Akgünoğlu. 2021. "Transfer and friction characteristics of sports socks fabrics made of synthetic fibres in different structures." *Tekstilec* 64 (4): 325–337.

Kim, Hey-sang, Stephen Michielsen, and Emiel DenHartog. 2021. "Wicking in textiles at rates comparable to human sweating." *Colloids and Surfaces A: Physicochemical and Engineering Aspects* 622:126726.

Kuklane, Kalev. 2004. "The use of footwear insulation values measured on a thermal foot model." *International Journal of Occupational Safety and Ergonomics* 10 (1):79–86.

Kumar, Bipin, Apurba Das, Ning Pan, R Alagirusamy, Rupali Gupta, and Jitender Singh. 2015. "Liquid transmission characteristics of padding bandages under pressure." *Journal of Biomaterials Applications* 30 (5):589–598.

Laing, Raechel M, Cheryl A Wilson, Linda A Dunn, and Brian E Niven. 2015. "Detection of fiber effects on skin health of the human foot." *Textile Research Journal* 85 (17):1849–1863.

Liu, Rong, Xia Guo, Terence T Lao, and Trevor Little. 2017. "A critical review on compression textiles for compression therapy: Textile-based compression interventions for chronic venous insufficiency." *Textile Research Journal* 87 (9):1121–1141.

Lou, Lun, Kaikai Chen, and Jintu Fan. 2021. "Advanced materials for personal thermal and moisture management of health care workers wearing PPE." *Materials Science and Engineering: R: Reports* 146:100639.

Manasoglu, Gizem, Rumeysa Celen, Mehmet Kanik, and Yusuf Ulcay. 2019. "Electrical resistivity and thermal conductivity properties of graphene-coated woven fabrics." *Journal of Applied Polymer Science* 136 (40):48024.

Manshahia, M, and A Das. 2013. "Comfort characteristics of knitted active sportswear: Heat and mass transfer." *Research Journal of Textile and Apparel* 17 (3):50–60.

Meyer-Heim, A, M Rothmaier, M Weder, J Kool, P Schenk, and J Kesselring. 2007. "Advanced lightweight cooling-garment technology: Functional improvements in thermosensitive patients with multiple sclerosis." *Multiple Sclerosis Journal* 13 (2):232–237.

Mokhtari Yazdi, Motahareh, and Mohammad Sheikhzadeh. 2014. "Personal cooling garments: A review." *The Journal of The Textile Institute* 105 (12):1231–1250.

Morrissey, Matthew P, and René M Rossi. 2014. "The effect of wind, body movement and garment adjustments on the effective thermal resistance of clothing with low and high air permeability insulation." *Textile Research Journal* 84 (6):583–592.

Najafi, Bijan, Hooman Mohseni, Gurtej S Grewal, Talal K Talal, Robert A Menzies, and David G Armstrong. 2017. "An optical-fiber-based smart textile (smart socks) to manage biomechanical risk factors associated with diabetic foot amputation." *Journal of Diabetes Science and Technology* 11 (4):668–677.

Narayana, Harishkumar, Jinlian Hu, Bipin Kumar, Songmin Shang, Jianping Han, Pengqing Liu, Tan Lin, FengLong Ji, and Yong Zhu. 2017. "Stress-memory polymeric filaments for advanced compression therapy." *Journal of Materials Chemistry B* 5 (10):1905–1916.

Nusser, Michaela, and Veit Senner. 2010. "High-tech-textiles in competition sports." *Procedia Engineering* 2 (2):2845–2850.

Parada, Marcelo, Peter Vontobel, René M Rossi, Dominique Derome, and Jan Carmeliet. 2017. "Dynamic wicking process in textiles." *Transport in Porous Media* 119 (3):611–632.

Priyalatha, S, and D Raja. 2017. "Investigation on wicking behavior of the knitted fabrics under different deformation state." *The Journal of The Textile Institute* 108 (7):1112–1121.

Psikuta, Agnes, Jonas Allegrini, Barbara Koelblen, Anna Bogdan, Simon Annaheim, Natividad Martínez, Dominique Derome, Jan Carmeliet, and René M Rossi. 2017. "Thermal manikins controlled by human thermoregulation models for energy efficiency and thermal comfort research–A review." *Renewable and Sustainable Energy Reviews* 78:1315–1330.

Purvis, Alison, and Helen Tunstall. 2004. "Effects of sock type on foot skin temperature and thermal demand during exercise." *Ergonomics* 47 (15):1657–1668.

Reyzelman, Alexander M, Chia-Ding Shih, Gregory Tovmassian, Mohan Nathan, Ran Ma, Henk Jan Scholten, Kara Malhotra, and David G Armstrong. 2022. "An evaluation of real-world smart sock–based temperature monitoring data as a physiological indicator of early diabetic foot injury: Case-control study." *JMIR Formative Research* 6 (4):e31870.

Roriz, Paulo, Susana Silva, Orlando Frazao, and Susana Novais. 2020. "Optical fiber temperature sensors and their biomedical applications." *Sensors* 20 (7):2113.

Rossi, Rene M, Rolf Stämpfli, Agnes Psikuta, Ivo Rechsteiner, and Paul A Brühwiler. 2011. "Transplanar and in-plane wicking effects in sock materials under pressure." *Textile Research Journal* 81 (15):1549–1558.

Scholten, Henk Jan, Chia-Ding Shih, Ran Ma, Kara Malhotra, and Alexander M Reyzelman. 2022. "Utilization of a smart sock for the remote monitoring of patients with peripheral neuropathy: Cross-sectional study of a real-world registry." *JMIR Formative Research* 6 (3):e32934.

Smith, C, and George Havenith. 2012. "Body mapping of sweating patterns in athletes: A sex comparison." *Medicine and Science in Sports and Exercise* (44):2350–2361.

Torreblanca González, José, Beatriz Gómez-Martín, Ascensión Hernández Encinas, Jesús Martín-Vaquero, Araceli Queiruga-Dios, and Alfonso Martínez-Nova. 2021. "The use of infrared thermography to develop and assess a wearable sock and monitor foot temperature in diabetic subjects." *Sensors* 21 (5):1821.

Van Roekel, Nick L, Erin M Poss, and David S Senchina. 2014. "Foot temperature during thirty minutes of treadmill running in cotton-based versus olefin-based athletic socks." *Bios* 85 (1):30–37.

Wang, Faming, S Annaheim, M Morrissey, and RM Rossi. 2014. "Real evaporative cooling efficiency of one-layer tight-fitting sportswear in a hot environment." *Scandinavian Journal of Medicine & Science in Sports* 24 (3):e129–e139.

Wang, Yong, Qifan Qiao, Zuowei Ding, and Fengxin Sun. 2021. "Strain-dependent wicking behavior of cotton/lycra elastic woven fabric for sportswear." *e-Polymers* 21 (1):263–271.

Wang, Yong, Qifan Qiao, Zongqian Wang, Changlong Li, and Stuart Gordon. 2022. "Exploring the transverse wicking behavior of mechanically robust warp super-elastic woven fabric for tight-fitting garments." *Textile Research Journal* 92 (9–10):1587–1597.

West, Anna M, George Havenith, and Simon Hodder. 2021. "Are running socks beneficial for comfort? The role of the sock and sock fiber type on shoe microclimate and subjective evaluations." *Textile Research Journal* 91 (15–16):1698–1712.

West, Anna Maria, Florian Oberst, James Tarrier, Christian Heyde, Heiko Schlarb, Gert-Peter Brüggemann, Simon Hodder, and George Havenith. 2020. "A thermal foot manikin as a tool for footwear evaluation and development." *Proceedings of the Institution of Mechanical Engineers, Part P: Journal of Sports Engineering and Technology*237 (1):1754337120952229.

Zhao, Shumi, Rong Liu, Xinbo Wu, Chongyang Ye, and Abdul Wasy Zia. 2020. "A programmable and self-adaptive dynamic pressure delivery and feedback system for efficient intermittent pneumatic compression therapy." *Sensors and Actuators A: Physical* 315:112285.

Zhu, Mengjia, Adrian Ferstera, Stejara Dinulescu, Nikolas Kastor, Max Linnander, Elliot W Hawkes, and Yon Visell. 2022. *A peristaltic soft, wearable robot for compression and massage therapy.* arXiv preprint arXiv:2206.01339.

8 Models and Simulations to Predict the Pressure on the Lower Limb Generated by Compression Devices

Nimesh Kankariya

CONTENTS

8.1 INTRODUCTION

Theoretical models define the behavior of a system using mathematical language and concepts. These mathematical languages and concepts are composed of constants, variables (known and unknown), and operators such as differential operators, algebraic operators, and functions. The operators establish the relationship between the output and the input of the system by means of constants and variables to formulate the equation (Willems and Polderman 2013). The mathematical model helps to predict the system behavior and compute the effectiveness of the factors under different conditions e.g. discrete vs continuous, static vs dynamic. The validity of the mathematical model depends on the presence and absence of various parameters, and consideration of the different variables and constants may change the outcome.

DOI: 10.1201/9781003298526-8

Therefore, understanding all parameters and the relationships among these parameters which affect the system is needed for an accurate mathematical model.

8.2 PRINCIPLES OF COMPRESSION THERAPY

Two major principles associated with compression therapy include the application of interface pressure which distributes the pressure evenly to an enclosed system and allows re-distribution of interface pressure according to the specific limb profile, and the application of materials with properties appropriate for use in a compression device (Fletcher, Moffatt, Partsch, Vowden et al. 2013). Controlling venous disease with compression generates some pressure on the interface between the skin and the compression device, the interface pressure (Mosti, Mattaliano, Polignano and Masina 2009). The muscles expand against the evenly distributed interface pressure throughout the limb, which generates the compressive effect. Laplace's law and Pascal's law are the underlying principles which describe how compression therapy delivers pressure to a limb. Laplace's law explains how the pressure is applied by a compression fabric, whereas Pascal's law explains the dynamics of compression pressure.

8.2.1 Pascal's Law

Pascal's law states that pressure is transmitted evenly when applied to an enclosed system of incompressible fluid. If the pressure at any point on an enclosed contained fluid increases, then the pressure at every other point in the container will increase equally. A capped tube of toothpaste with multiple equally sized holes punched in it can be used to demonstrate this law. When pressure is applied to the tube at one point, toothpaste will outflow at the same rate from all the openings, regardless of their distance from the applied pressure (Fletcher et al. 2013). The soft tissue in the human body resembles an incompressible fluid (Dubuis, Avril, Debayle and Badel 2012). According to Pascal's law, the external pressure applied by compression stocking to a body part forms an enclosed container, and the pressure is thus further distributed to the underlying soft tissues equally in all directions. Compression bandage layers, with a high static stiffness index, create a rigid cylinder when applied to a lower limb, producing higher recoil forces which squeeze the internal soft tissue and result in a narrowing of veins diameter and improved hemodynamic effectiveness of the lower extremities (Partsch 2005, Schuren and Mohr 2010, Fletcher et al. 2013).

8.2.2 Laplace's Law and Modification

Laplace's law, deduced by Young and Laplace (1805), described a principle that defined the pressures applied to curved surfaces (Pellicer, García-Morales and Hernåndez 2000). According to Laplace's law, the pressure difference across a closed sphere vessel is directly proportional to the corresponding tension produced in the sphere and is given by equation.

$$P\alpha - P\beta = \frac{2T}{r} \tag{8.1}$$

where $P\alpha$ and $P\beta$ are internal and external pressure of the closed spherical surface, respectively (Pa), r is the radius of the curvature (m), and T is the tension generated on the vessel wall (N/m).

The pressure applied inside is always greater than the pressure applied outside of a spherical surface (Equation 8.1). The difference between inside and outside pressures tends to zero as the radius becomes infinite. In contrast, the difference tends to infinity when the radius of the curvature (r) reaches zero. However, practically this case does not arise. For a given vessel radius and internal pressure, a cylindrical vessel will have two times the wall tension of a spherical vessel. While calculating the pressures (P) in the wall of the cylindrical vessel (blood vessel), the equation needs to be modified (Equation 8.2) (Thomas 2003, Nave Accessed on 7 Jan 2022). According to Equation 8.2 the larger the vessel radius, the larger the wall tension required to withstand a given internal fluid pressure.

$$P = \frac{T}{r} \tag{8.2}$$

where P is pressure (Pa), r is the radius of the curvature (m), and T is tension (N/m).

Many researchers have proposed numerous modified forms of Laplace's law to calculate the interface pressure between fabric and skin (Kirk and Ibrahim 1966, Hui and Ng 2001, Macintyre 2007, Seo, Kim, Cordier and Hong 2007, Yıldız 2007, Valentinuzzi and Kohen 2011, Gaied, Drapier and Lun 2006, Thomas 2003). When vessel wall thickness (t) is taken into account, then Equation 8.2 can be revised to form Equation 8.3 (Valentinuzzi and Kohen 2011). When the number of layers and width of the bandage are considered, then Equation 8.2 can be modified to form Equation 8.4 (Thomas 2003).

$$P = \frac{T \times t}{r} \tag{8.3}$$

$$P = \frac{T \times n}{W \times r} \tag{8.4}$$

where P is pressure (Pa), r is the radius of the curvature (m), T is tension (N/m), t is thickness (m), W is the width of the bandage (m), and n is the number of layers.

The pressure generated on the human body is a function of tension applied by the pressure fabric. Compression stockings or garments have circumferences smaller than the body circumference on which these are applied so that the fabric extends when applied and remains stretched during wear. Bandages are also applied with tension. The tension in the fabric results in an inward force over the curved human body parts. Textile compression devices, e.g. bandages and stockings, resemble a cylinder and produce pressure on the human body. Therefore, textile properties i.e. mechanical properties including fabric thickness play a vital role in the application of pressure and the extent and consistency of pressure exerted. To calculate the interface pressure exerted by thin and thick fabric, Equations 8.2 and 8.3 respectively can be used. For modeling purposes, if the thicknesses of fabrics used in compression therapy are assumed to be negligible, then Equation 8.2 can be used.

The pressure exerted by the compression textile is proportional to the tension generated in the fabric (Equation 8.2). The tension in the fabric is proportional to the elongation (level of stretch) of the fabric and the relationship is not linear (Thomas and Fram 2003). In compression intervention, a wide range of treatment pressures is used, which is achieved by controlling the level of stretch, with an accuracy of application largely affected by practitioner skill and experience (Hafner, Lüthi, Hänssle, Kammerlander et al. 2000, Moffatt 2008). In a comparative study of four-layer bandage vs short stretch bandage, Partsch et al. (2001) concluded that the differences in the outcome were attributed to the practitioner's skill in different centers with inelastic bandage systems. Some commercially available compression bandages (i.e. DYNA-FLEX™ multilayer compression system, Surepress™) include geometric patterns printed on them which change from one shape to the other (e.g. oval-shape to circle-shape, rectangular-shape to square-shape) at the correct level of stretch, which assists practitioners in achieving consistent and reproducible pressure (Thomas and Fram 2003).

The curvature of the body differs from one place to another. It is challenging to determine exactly the pressure exerted in complex-shaped body areas of the calf, ankle, and knee. As the knee bends, the pressure applied by the compression fabric distributes into multi-axial directions. The tension (tensile stress) produced in fabric and the radius of curvature of a specific bending position in horizontal and vertical directions was measured and substituted into the relationship reported by Kirk and Ibrahim (1966) to calculate the degree of pressure applied (Equation 8.5).

$$P = \frac{T_H}{r_H} + \frac{Tv}{rv} \tag{8.5}$$

where P is pressure (lb/in^2), r_H is the radius of curvature in horizontal directions (in), r_V is the radius of curvature in vertical directions (in), T_H is tension in the horizontal direction (lb/in), and T_V is tension in the vertical direction (lb/in).

To simulate the clothing pressure, Equation 8.5 was proposed by Seo et al. (2007) in which the fabric thickness was accommodated in order to predict the pressure at various body parts with complex curvature.

$$P = \frac{\sigma T_H \times t}{r_H} + \frac{\sigma_{TV} \times t}{r_V} \tag{8.6}$$

where P is pressure (gf/cm^2), r_H is the radius of curvature in horizontal directions (cm), r_V is the radius of curvature in vertical directions (cm), σ_{TH} is tensile stress or tension force per unit area in the horizontal direction (gf/cm^2), σ_{TV} is tensile stress or tension force per unit area in the vertical direction (gf/cm^2), and t is the thickness (cm).

To obtain the value of pressure applied by multilayered fabrics, Hui and Ng (2001) modified equation 8.2 to accommodate the compression applied by the multilayer fabric and the relationship of this with the thickness of each individual layer. The pressure applied by a multilayer fabric was dependent on the thickness of each layer, total strain, elastic modulus, circumference of the body, and the number of layers (Equation 8.7).

$$P = \frac{\varepsilon\, 2\, \pi \sum_{1}^{n} (Ei\ ti)}{C} \tag{8.7}$$

where P is pressure (kPa), ε is the total strain (%), E is elastic modulus (N/m^2), t is thickness (m), c is the circumference of the body (m), and n is the number of layers.

However, while Laplace's equation provided a mechanistic view of the pressure applied on a rigid body, results obtained on the compressible human body may not be accurate (Basford 2002). This was attributed to difficulty assessing the radius of curvature of the human body. Additionally, when patients are measured for pressure fabrics in clinics or in hospitals, instead of radius, the circumference of the body part is measured. As a result, Macintyre (2007) proposed a modified Laplace's law to predict pressure from the circumference of the body part and fabric tension (Equation 8.8).

$$P = \frac{T \times 20000\, \pi}{C} \tag{8.8}$$

where P is pressure (Pa), c is the circumference of cylinder or body (m), and T is tension (N/m).

Macintyre (2007) also proposed using mmHg as the units of pressure measurement, which is the most commonly used unit for measurement of capillary pressure, instead of Pascal, which was found to be more suitable for calculating the pressure in mmHg generated on the human body (Equation 8.9). (SI unit of pressure is N/m^2, where 133.32 N/m^2 = 133.32 Pa = 1 mmHg)

$$P = \frac{T \times 471.3}{C} \tag{8.9}$$

where P is pressure (mmHg), c is the circumference of cylinder or body (m), and T is tension (N/m).

For the bandaging system, while calculating the pressure, the number of layers applied and the bandage width should also be taken into account with the same theory. For each stress value with predetermined bandage width and circumference values, one can get the theoretical pressures from Equation 8.10 (Yıldız 2007, Schuren and Mohr 2008).

$$P = \frac{T \times 4620 \times 10000 \times n}{C \times W} \tag{8.10}$$

where P is pressure (mmHg), c is the circumference of cylinder or body (m), T is tension (kgF), W is the width of the bandage (m), and n is the number of layers.

The ratio of the tension to the circumference (T/C) allows the fit of the fabric on the body, i.e. tight fit or loose, to be expressed (Troynikov, Ashayeri, Burton, Subic et al. 2010). A high T/C ratio indicates a tight-fitting fabric and a low T/C ratio a loose-fitting fabric. Measurement of T/Cratio helps in the design of compression stockings. Simply multiplying the number of layers as shown in Equation 8.10 suggests that the pressure produced is directly proportional to the number of layers, which is not correct. By comparing four different four-layer bandages, Dale et al. (2004) concluded that the final pressure obtained by a multilayer bandage was not cumulative. The final pressure achieved by a multilayer bandage was 70% of the pressure exerted by every single layer. Simply multiplying pressure by the number of layers was also questioned by Wertheim et al. (1999). Therefore, the use of Laplace's law to predict the sub-bandage pressure in multilayer bandaging system is questionable.

Based on thick wall cylinder theory a mathematical model was developed to determine the interface pressure produced by multilayer wrapping systems (Equation 8.11) (Khaburi, Nelson, Hutchinson and Dehghani-Sanij 2011). The model attempted to accommodate the thickness of the bandage. The model showed that the consequences of ignoring bandage thickness were an error of 19% or more when predicting the overall sub-bandage pressure produced by multilayer, multicomponent systems (Khaburi, Dehghani-Sanij, Nelson and Hutchinson 2012).

$$P = \sum_{i=1}^{n} \frac{T[dia+t+2t(i-1)]}{\left(\left(\frac{1}{2}\right) W\left[dia+2t(i-1)\right]^2\right)+(Wt\left[dia+t+2t(i-1)\right])} \times 0.0075 \tag{8.11}$$

where P is pressure (mmHg), dia is limb diameter (m), T is tension (N), W is the width of bandage (m), t is the thickness of bandage (m), and n is the number of layers.

The thickness of the bandage was assumed as constant for each successive layer added. However, due to the compressive forces of successive layers, the thickness of the bandage would not be constant. A further disadvantage of this model was that it was developed for a cylindrical limb geometry, when the human leg has a conical-shaped geometry (Pourazadi, Ahmadi and Menon 2015). Sikka et al. (2016) developed a mathematical model to predict the pressure on a conical-shaped lower leg. The model was developed for multilayer bandaging (Equation 8.12).

$$P = \frac{2\,T\sqrt{w^2-(r_2-r_1)^2}}{W\,dia\,w} \tag{8.12}$$

where P is pressure (mmHg), dia is limb diameter (m), T is tension (N), W is the width of bandage (m), w is the distance between two extreme points of bandage layers (m), and r_1, r_2 are the radius of curvature of the limb at smaller and bigger circumference of the limb (m).

Sikka et al. (2016) concluded that the thickness of the bandage and the geometry of the limb have a significant effect on the pressure produced by multilayer bandaging. The effect of interactions among bandaging techniques and changes in the radius of curvature of the leg in dynamic posture i.e. walking, running, or jogging over time were not considered in this study. Further research is needed to calculate the potential change in sub-bandage pressure induced by changes in limb shape/size as a result of calf muscle activity.

Cylindrical-shaped limb model developed by Khaburi et al. (2011) and conical-shaped limb model developed by Sikka et al. (2016) were the limited studies to predict the pressure applied by multilayer bandaging systems and have not been verified when applied to predict the pressure exerted by compression stockings. The effect of processing parameters (i.e. loop length, yarn count, yarn feeding tension) on the pressure performance of compression stockings was studied (Zhang, Sun, Li, Chen et al. 2019). Moreover, two theoretical models, while considering the lower leg as cylindrical- (Equation 8.13) and conical-shaped (Equation 8.14), were employed to investigate the pressure behaviors of compression stockings. The theoretical model for a conical-shaped lower leg predicted the pressure more accurately than the prediction by the cylinder shape.

$$P_i = \frac{N_{\theta i} t 2\pi}{C_i} \tag{8.13}$$

where P_i is pressure exerted on part i of the limb (kPa), t is the thickness of compression stocking (mm), C_i is the girth of part i (m), and $N_{\theta\ i}$ is the circumferential stress on part i of the limb.

$$P_i = \frac{T_i\sqrt{l^2 - (r_{i+1} - r_i)^2}}{r_{i+1}\ n\ l^2} \tag{8.14}$$

where P_i is the pressure exerted on part i of the limb (kPa), T_i is tension in the fabric of part i (m), n is the number of conical parts, l is the distance between neighboring circles (m), and r_{i+1}, r_i are the radius of curvature of the limb at bigger and smaller parts circumference of the limb (m).

Teyeme et al. (2021) developed a theoretical model to predict the compression pressure exerted by knit-based compression garments (Equation 8.15).

$$P = \frac{t \times 2\pi}{C}\left(A_1\left(\frac{C_f - C_i}{C_i}\right)^3 + A_2\left(\frac{C_f - C_i}{C_i}\right)^2 + A_3\left(\frac{C_f - C_i}{C_i}\right)^1 + B\right) \tag{8.15}$$

where P is pressure (Pa), C is the circumference of the cylinder when the fabric is wrapped around it (m), t is fabric thickness (m), C_f is the initial test fabric circumferences before stretching (m), C_i is the final test fabric circumference after stretching (m), A_1, A_2, A_3 are coefficient, and B is a constant.

The properties of the knit fabric that might influence the interface pressure, i.e. strain, modulus of elasticity, stress, and thickness, were measured and integrated into Laplace's law (Teyeme et al. 2021). It was concluded that the knit fabric with a greater compact structure and higher areal density led to more pressure than the loose-knit structures and knit structure with lower areal density. The differences in pressure levels between predicted and experimental values ranged from 0.06% to 9.42%. However, this study only considered a restricted number of cylinder circumferences and fabric types as limitations. A larger number of textiles and a broader range of circumferences should be included in subsequent analysis. The contact pressure was measured using a rigid cylinder method, but the human body (i.e. lower leg) is not rigid and has anatomical variation. The pressure applied by a garment with a certain tension is not uniform and is distributed variably throughout the various portions of the lower leg due to the non-uniformity shape of the lower leg.

8.3 MODELS FOR PNEUMATIC COMPRESSION DEVICE

An accurate model of the pressure generated by a pneumatic compression device (PCD) plays a crucial role in understanding the functional parameters and physical and mechanical behavior of the compression system. All variables which may affect the performance of a pneumatic device must be taken into account or controlled to obtain the expected outcome. The actual pressure exerted by PCD depends on various factors such as material properties, cross-sectional shape of the lower limb, inflation–deflation cycles, and extent of tissue deformation. The expected outcome for the given therapy is the replication of the action of a muscle pump irrespective of the physical movement of the patient (i.e. whether mobile or immobile). To reproduce the effect of a muscle pump and restore venous return, the generated pressure must be mechanically or robotically controlled with different setting conditions.

A model of a compression system, where multiple air chambers arranged in a sequential manner are applied on the lower leg, is illustrated in Figure 8.1. The air is supplied to some form of the pump (electrical or mechanical) through the inlet source, and the outlet pressurized air is subsequently delivered to the pneumatic bladders at various flow rates via a multiple-valve system controlled by a valve regulator. The pneumatic bladders, Bladder *i* (i.e. Bladder 1, Bladder 2, Bladder 3), connected to the sleeve are inflated by the air flow rate of $Q_{\text{bladder } i}$ (i.e. $Q_{\text{bladder 1}}$, $Q_{\text{bladder 2}}$, and $Q_{\text{bladder 3}}$, respectively), resulting in the production of a bladder pressure of $P_{\text{bladder } i}$ (i.e. $P_{\text{bladder 1}}$, $P_{\text{bladder 2}}$, and $P_{\text{bladder 3}}$, respectively). When these inflated air bladders come into contact with the associated positions on the lower leg i.e. ankle, calf, below knee, they produce corresponding contact pressures $P_{\text{leg i}}$ (i.e. $P_{\text{leg 1}}$, $P_{\text{leg 2}}$, and $P_{\text{leg 3}}$, respectively) in their separate region of operation. In the contact condition of air bladders with the leg, the leg tissue deforms (Z_{leg}). Following the inflation phase, the air is released via an exhaust valve and bladders deflate. In a clinical context, the bladders inflate and deflate intermittently to generate the desired pressure on the lower limb. The pressure exerted cyclically (i.e. inflation and deflation times) varies among pneumatic devices (Rithalia, Heath and Gonsalkorale 2002). To regulate the inflation/deflation time, and therefore to apply the required pressure pattern on the lower limb,

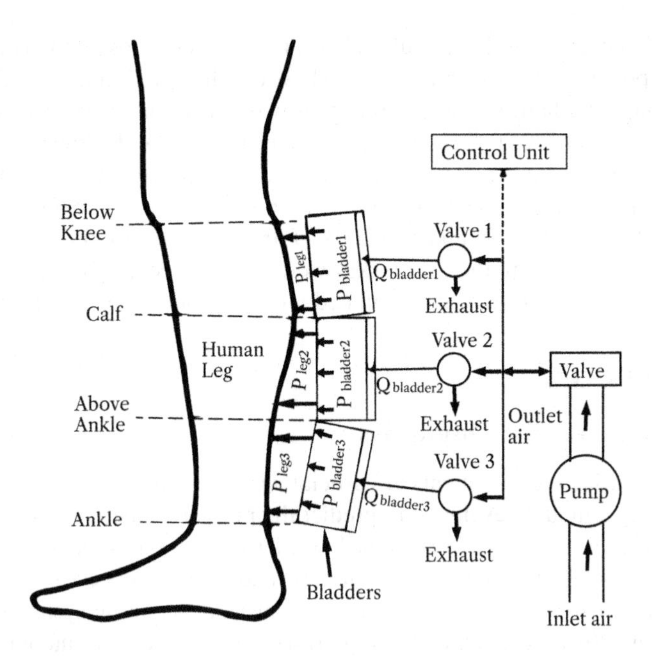

FIGURE 8.1 Model of pneumatic compression device with multiple pressure chambers, when pressure is applied to lower leg.

PID (Proportional, Integral, Derivative) algorithm-based control technique was used (Ferraresi, Maffiodo and Hajimirzaalian 2014). An advanced programmable online monitoring system, and self-adaptive controlled dynamic pressure delivery system, was developed by Zhao et al. (2020) to enhance the effectiveness of PCDs.

A theoretical model to predict the pressure applied by the pneumatic device on a regular shaped lower leg was developed (Ferraresi, Maffiodo and Hajimirzaalian 2014, Zhao, Liu, Wu, Ye et al. 2020). The model was validated using simulation by MATLAB® software (Ferraresi, Maffiodo and Hajimirzaalian 2014), but not validated experimentally, nor has the model been checked against models of varying radii. Based on flow continuity theory a mathematical model was developed by Kankariya et al. (2021a) to predict the pressure inside the Bladder i ($P_{\text{bladder } i}$) which is wrapped around the irregular-shaped lower leg (i.e. circular in shape, elliptic in shape) and consequentially to predict the bladder-skin contact pressure ($P_{\text{leg } i}$). The models to predict the pressure applied by the compression sleeves on the circular- or elliptical-shaped lower leg segments were compared against experimental data for three different positions (i.e. ankle to above ankle, above ankle to below the calf, below the calf to below the knee) on the lower leg manikin. Properties of silicone material were measured (Kankariya, Laing and Wilson 2020, Kankariya, Wilson and Laing 2021), and compression sleeves were fabricated (Kankariya, Laing and Wilson 2021a). In comparison to the model developed for an elliptical shape of the lower leg, the circular-shaped lower leg manikin predicted the pressure more accurately for position below the calf to below the knee, whereas the model developed for an elliptical-shaped lower leg calculated the pressure more closely to the experimental data for position ankle to above ankle. However, the manikin limb utilized in this investigation had a hard surface texture that was consistent all the way around the lower leg. In reality, the surface texture (tissue properties) of the lower extremities at any cross-section is dissimilar, so the deformation of tissues in a human leg at any cross-section will be irregular when the compression device is applied (Kankariya, Laing and Wilson 2021b, Kankariya 2022). Further, the physical dimensions of the pressure sensor including the calico skin were not considered, and the thickness of the pressure sensor increases when inflated with air, potentially affecting results.

8.4 LOWER LEG

8.4.1 Landmarks and Dimensions

Accurate lower leg measurements are crucial when designing a compression device and validating a model developed to predict the pressure generated by compression devices. Some researchers measured the lower leg dimensions manually using simple measuring tools i.e. tape and Vernier calipers, and some other researchers collected these dimensions indirectly using a three-dimensional scanning system (Gong, Fei, Lai and Liang 2010, Lee, Lin and Wang 2014). The accuracy of outcomes obtained from the manual approach is affected by human error. The three-dimensional scanning method is the most recommended approach for measuring lower leg dimensions due to higher precision and accuracy. However, three-dimensional methods have the disadvantage of very high initial set-up cost (Lee, Lin and Wang 2014).

Among various lower leg dimensions, calf, below knee, and ankle circumference are the most important dimensions for sizing and fitting of the compression device to the lower leg. However, the use of other dimensions, i.e. distance from below knee to calf, distance from calf to ankle, and distance from ankle to below knee, precisely controls the fit and shape of the device due to their close relationship with other critical leg dimensions. Landmarks (i.e. bony prominences at the ankle region, maximum circumference between the ankle and the knee (denoted as calf), minimum circumference between the calf and the knee (denoted as below the knee)) present on the lower limb, and descriptions of the lower leg are presented in Figure 8.2 (Table 8.1) (Granberry et al. 2017, International Organization for Standardization 2017).

8.4.2 Tissue Properties and Tissue Deformation

The lower leg is the structure between the knee and the ankle and consists of many different tissues. The pressure exerted varies due to the diverse properties of the tissue and their relative proportions at each given cross-section (i.e. hard tissues of bony parts at tibial crest position; soft tissue of muscular parts at calf position) (Giele et al. 1997). The behavior of the tissues of the lower leg when exposed to a load (as in a pressure) is described by its deformation and stiffness. Deformation refers to the shape and volume change of an object as a result of pressure being applied. Stiffness denotes the rigidity of the material which resists deformation caused by an external applied

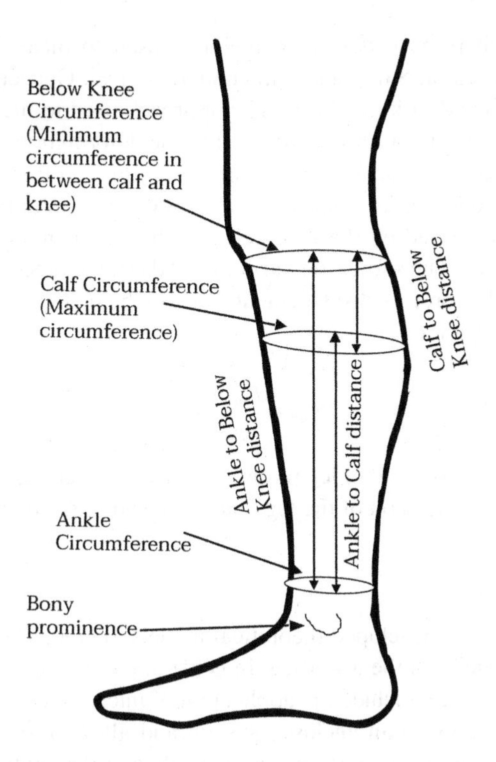

FIGURE 8.2 Position for defining lower leg dimensions.

TABLE 8.1
Definitions of Lower Limb Positions

Dimension	Definition
Calf circumference	Calf circumference is located at the apex of the gastrocnemius muscle
Below knee circumference	Below knee circumference is located at the bottom-most part of the patella
Ankle circumference	Ankle circumference is located at the narrowest part of the leg just above the medial malleolus
Below knee to ankle distance	Distance from below knee to ankle
Calf to below knee distance	Distance from calf to below knee
Ankle to calf distance	Distance from ankle to calf

force. The concept of stiffness follows Hooke's law which states that within certain limits the force required to stretch or compress the material is directly proportional to the displacement of the material and the stiffness constant. In terms of a human leg, stiffness is the connection between the applied force and deformation of the leg tissue. More induced strain might be expected in soft tissue than in harder tissue.

Various researchers have described methods used to measure leg tissue stiffness (Horikawa, Ebihara, Sakai and Akiyama 1993, Dai, Gertler and Kamm 1999, Muraki, Fukumoto and Fukuda 2013). The approach used to measure the stiffness is to apply a known value of force and measure the deformation of tissue caused by that force. The force terminal is used to apply the force on the surface, and this is attached to a rod and load cell. A spring absorbs the vibration of the rod. The known amount of force is applied by the load cell and displacement is measured via displacement transducer. Therefore, the variation in the force determines the variation in displacement. The force vs displacement can then be plotted, and stiffness can be calculated from the plot as:

$$K_{leg} = \frac{\Delta F_{tisue\_deform}}{\Delta Z_{leg}} \quad (8.16)$$

where K_{leg} is stiffness constant of leg tissue, ΔF_{tissue_deform} is change in force, and ΔZ_{leg} is the change in the dimension of the leg due to change in force applied.

8.5 CONCLUSION

Several researchers had developed theoretical models to predict the pressure applied by compression devices on the lower leg. In general, two major principles associated with compression therapy include the application of interface pressure which distributes the pressure evenly to an enclosed system and allows re-distribution of interface pressure according to the specific limb profile and the application of materials with properties appropriate for use in a compression device. However, the theoretical models developed in most of the studies were validated using a hard surface regular-shaped lower leg manikin. The pressure exerted by the compression device is affected by the properties of the underlying tissue at any particular cross-section of the human lower leg (i.e. soft tissue of muscular parts, hard tissues of bony parts). Moreover, the shape at any portion/segment of the lower leg and radius of curvature at any cross-section of the lower leg (i.e. calf, tibial crest) is irregular and differs among individuals and populations. Hence, the theoretical models for the human lower leg with irregular shape and tissue properties should further be validated using a human leg.

ACKNOWLEDGMENTS

The author would like to thank Prof. Raechel M Laing and Prof. Cheryl A Wilson for their support.

REFERENCES

Basford, J R. 2002. "The law of Laplace and its relevance to contemporary medicine and rehabilitation." *Archives of Physical Medicine and Rehabilitation* 83 (8):1165–1170. doi:10.1053/apmr.2002.33985.

Dai, G, Gertler, J P, and Kamm, R D. 1999. "The effects of external compression on venous blood flow and tissue deformation in the lower leg." *Journal of Biomechanical Engineering* 121 (6):557–564. doi:10.1115/1.2800853.

Dale, J J, Ruckley, C V, Gibson, B, Brown, D, Lee, A J, and Prescott, R J. 2004. "Multilayer compression: Comparison of four different four-layer bandage systems applied to the leg." *European Journal of Vascular and Endovascular Surgery* 27 (1):94–99. doi:10.1016/j.ejvs.2003.10.014.

Dubuis, L, Avril, S, Debayle, J, and Badel, P. 2012. "Identification of the material parameters of soft tissues in the compressed leg." *Computer Methods in Biomechanics and Biomedical Engineering* 15 (1):3–11. doi:10.1080/10255842.2011.560666.

Ferraresi, C, Maffiodo, D, and Hajimirzaalian, H. 2014. "A model-based method for the design of intermittent pneumatic compression systems acting on humans." *Proceedings of the Institution of Mechanical Engineers, Part H: Journal of Engineering in Medicine* 228 (2):118–126. doi:10.1177/0954411913516307.

Fletcher, J, Moffatt, C, Partsch, H, Vowden, K, and Vowden, P. 2013. "Principles of compression in venous disease: A practitioner's guide to treatment and prevention of venous leg ulcers." *Wounds International* 2013:1–21.

Gaied, I, Drapier, S, and Lun, B. 2006. "Experimental assessment and analytical 2D predictions of the stocking pressures induced on a model leg by medical compressive stockings." *Journal of Biomechanics* 39 (16):3017–3025. doi:10.1016/j.jbiomech.2005.10.022.

Giele, H P, Liddiard, K, Currie, K, and Wood, F M. 1997. "Direct measurement of cutaneous pressures generated by pressure garments." *Burns* 23 (2):137–141. doi:10.1016/s0305-4179(96)00088-5.

Gong, T, Fei, R, Lai, J, and Liang, G. 2010. "Foot shape analysis of adult male in the China." In *Information computing and applications*, edited by Rongbo Zhu, Yanchun Zhang, Baoxiang Liu and Chunfeng Liu, 1–7. Berlin, Heidelberg: Springer Berlin Heidelberg. doi:10.1007/978-3-642-16339-5_1.

Granberry, R, Duvall, J, Dunne, L E, and Holschuh, B. 2017. "An analysis of anthropometric geometric variability of the lower leg for the fit & function of advanced functional garments." *Proceedings of the 2017 ACM International Symposium on Wearable Computers*, Maui, Hawaii, USA.

Hafner, J, Lüthi, W, Hänssle, H, Kammerlander, G, and Burg, G. 2000. "Instruction of compression therapy by means of interface pressure measurement." *Dermatologic Surgery* 26 (5):481–488. doi:10.1046/j.1524-4725.2000.99257.x.

Horikawa, M, Ebihara, S, Sakai, F, and Akiyama, M. 1993. "Non-invasive measurement method for hardness in muscular tissues." *Medical & Biological Engineering & Computing* 31 (6):623–627. doi:10.1007/BF02441811.

Hui, C L, and Ng, S F. 2001. "Model to predict interfacial pressures in multilayer elastic fabric tubes." *Textile Research Journal* 71 (8):683–687. doi:10.1177/004051750107100806.

International Organization for Standardization. 2017. ISO 8559-1. In *Size designation of clothes – Part 1: Anthropometric definitions for body measurement*. Geneva: International Organization for Standardization.

Kankariya, N. 2022. "Material, structure, and design of textile-based compression devices for managing chronic edema." *Journal of Industrial Textiles* 12 (2022):1–35. doi:10.1177/15280837221118844.

Kankariya, N, Laing, R M, and Wilson, C A. 2021a. "Prediction of applied pressure on model lower limb exerted by an air pneumatic device." *Medical Engineering & Physics* 97 (2021):77–87. doi:10.1016/j.medengphy.2021.07.007. Copyright (2021). Elsevier Publication.

Kankariya, N, Laing, R M, and Wilson, C A. 2021b. "Textile-based compression therapy in managing chronic oedema: Complex interactions." *Phlebology* 36 (2):100–113. doi:10.1177/0268355520947291.

Kankariya, N, Laing, R M, and Wilson, C A. 2020. "Challenges in characterising wool knit fabric component of a textile based compression intervention." International Virtu-Wool Research Conference – 26–27 Aug, 2020, New Zealand, AgResearch.

Kankariya, N, Wilson, C A, and Laing, R M. 2021. "Thermal and moisture behavior of a multi-layered assembly in a pneumatic compression device." *Textile Research Journal* 92 (15–16):2669–2684. doi:10.1177%2F00405175211006942.

Khaburi, J A, Dehghani-Sanij, A A, Nelson, E A, and Hutchinson, J. 2012. "Effect of bandage thickness on interface pressure applied by compression bandages." *Medical Engineering & Physics* 34 (3):378–385. doi:10.1016/j.medengphy.2011.07.028.

Khaburi, J A, Nelson, E A, Hutchinson, J, and Dehghani-Sanij, A A. 2011. "Impact of multilayered compression bandages on sub-bandage interface pressure: A model." *Phlebology* 26 (2):75–83. doi:10.1258/phleb.2010.009081.

Kirk, J W, and Ibrahim, S M. 1966. "Fundamental relationship of fabric extensibility to anthropometric requirements and garment performance." *Textile Research Journal* 36 (1):37–47. doi:10.1177/004051756603600105.

Lee, Y C, Lin, G, and Wang, M J. 2014. "Comparing 3D foot scanning with conventional measurement methods." *Journal of Foot and Ankle Research* 7 (1):44–54. doi:10.1186/s13047-014-0044-7.

Macintyre, L. 2007. "Designing pressure garments capable of exerting specific pressures on limbs." *Burns* 33 (5):579–586. doi:10.1016/j.burns.2006.10.004.

Moffatt, C. 2008. "Variability of pressure provided by sustained compression." *International Wound Journal* 5 (2):259–265. doi:10.1111/j.1742-481x.2008.00470.x.

Mosti, G, Mattaliano, V, Polignano, R, and Masina, M. 2009. "Compression therapy in the treatment of leg ulcers." *Acta Vulnologica*:113–135.

Muraki, S, Fukumoto, K, and Fukuda, O. 2013. "Prediction of the muscle strength by the muscle thickness and hardness using ultrasound muscle hardness meter." *SpringerPlus* 2 (457):1–7. doi:10.1186/2193-1801-2-457.

Nave, C R. "Laplace's law." Georgia State University, accessed 7/01/2022. http://hyperphysics.phy-astr.gsu.edu/hbase/ptens.html.

Partsch, H. 2005. "The static stiffness index: A simple method to assess the elastic property of compression material in vivo." *Dermatologic Surgery* 31 (6):625–630. doi:10.1111/j.1524-4725.2005.31604.

Partsch, H, Damstra, R J, Tazelaar, D J, Schuller-Petrovic, S, Velders, A J, De Rooij, M J, Sang, R R, and Quinlan, D. 2001. "Multicentre, randomised controlled trial of four-layer bandaging versus short-stretch bandaging in the treatment of venous leg ulcers." *VASA* 30 (2):108–113. doi:10.1024/0301-1526.30.2.108.

Pellicer, J, García-Morales, V, and Hernández, M J. 2000. "On the demonstration of the Young-Laplace equation in introductory physics courses." *Physics Education* 35 (2):126–129. doi:10.1088/0031-9120/35/2/309.

Pourazadi, S, Ahmadi, S, and Menon, C. 2015. "On the design of a DEA-based device to potentially assist lower leg disorders: An analytical and fem investigation accounting for nonlinearities of the leg and device deformations." *Biomedical Engineering Online* 14 (1):1–18. doi:10.1186/s12938-015-0088-3.

Rithalia, S V S, Heath, G H, and Gonsalkorale, M. 2002. "Evaluation of intermittent pneumatic compression systems." *Journal of Tissue Viability* 12 (2):52–57. doi:10.1016/S0965-206X(02)80014-6.

Schuren, J, and Mohr, K. 2008. "The efficacy of Laplace's equation in calculating bandage pressure in venous leg ulcers." *Wounds* 4 (2):38–47.

Schuren, J, and Mohr, K. 2010. "Pascal's law and the dynamics of compression therapy: A study on healthy volunteers." *International Angiology* 29 (5):431–435. PMID: 20924347.

Seo, H, Kim, S J, Cordier, F, and Hong, K. 2007. "Validating a cloth simulator for measuring tight-fit clothing pressure." *Proceedings of the 2007 ACM symposium on Solid and Physical Modeling*, Beijing, China.

Sikka, M P, Ghosh, S, and Mukhopadhyay, A. 2016. "Mathematical modeling to predict the sub-bandage pressure on a conical limb for multi-layer bandaging." *Medical Engineering & Physics* 38 (9):917–921. doi:10.1016/j.medengphy.2016.05.006.

Teyeme, Y, Malengier, B, Tesfaye, T, Vasile, S, Endalew, W, and Van, L L. 2021. "Predicting compression pressure of knitted fabric using a modified Laplace's law." *Materials* 14 (16-4461):1–13. doi:10.3390/ma14164461.

Thomas, S. 2003. "The use of the Laplace equation in the calculation of sub-bandage pressure." *European Wound Management Association* 3 (1):21–23.

Thomas, S, and Fram, P. 2003. "Laboratory-based evaluation of a compression-bandaging system." *Nursing Times* 99 (40):24–28. PMID: 14603620.

Troynikov, O, Ashayeri, E, Burton, M, Subic, A, Alam, F, and Marteau, S. 2010. "Factors influencing the effectiveness of compression garments used in sports." *Procedia Engineering* 2 (2):2823–2829. doi:10.1016/j.proeng.2010.04.073.

Valentinuzzi, M E, and Kohen, A K. 2011. "Laplace's law: What it is about, where it comes from, and how it is often applied in physiology." *IEEE Pulse* 2 (4):74–84. doi:10.1109/MPUL.2011.942054.

Wertheim, D, Melhuish, J, Williams, R, and Harding, K. 1999. "Measurement of forces associated with compression therapy." *Medical & Biological Engineering & Computing* 37 (1):31–34. doi:10.1007/bf02513262.

Willems, J C, and Polderman, J W. 2013. *Introduction to mathematical systems theory: A behavioral approach*. New York: Springer Science & Business Media.

Yıldız, N. 2007. "A novel technique to determine pressure in pressure garments for hypertrophic burn scars and comfort properties." *Burns* 33 (1):59–64. doi:10.1016/j.burns.2006.04.026.

Zhang, L, Sun, G, Li, J, Chen, Y, Chen, X, Gao, W, and Hu, W. 2019. "The structure and pressure characteristics of graduated compression stockings: Experimental and numerical study." *Textile Research Journal* 89 (23–24):5218–5225. doi:10.1177/0040517519855319.

Zhao, S, Liu, R, Wu, X, Ye, C, and Zia, A W. 2020. "A programmable and self-adaptive dynamic pressure delivery and feedback system for efficient intermittent pneumatic compression therapy." *Sensors and Actuators A: Physical*:112285. doi:10.1016/j.sna.2020.112285.

9 Pressure Sensors to Measure Interface-Pressure

Martin Camenzind, Rolf Stämpfli, René M. Rossi

CONTENTS

9.1 INTRODUCTION

In medical applications and the case of integration of sensors into garments and especially compression garments, defined skin contact and the possibility to assess interface pressure are essential. Flexible foils or textiles based on specific fibers and adapted structures can be used as a substrate for sensing elements ensuring good contact due to their flexibility.

Typical applications for flexible sensors to measure contact pressure are e-skin (e.g. robotics), body monitoring (e.g. heart rate or breathing frequency), detection of sleeping patterns (e.g. body posture), or other medical assessments (e.g. gait cycle assessment with sensors in shoes).

DOI: 10.1201/9781003298526-9

Regarding working principles, flexible pressure sensors are usually divided into four groups: resistivity, capacitance, piezoelectricity, or other working mechanisms (e.g. optical effects, triboelectricity, magnetoresistivity, pyroelectricity, thermoelectricity, or other physical effects) (Chen and Yan 2020).

The majority of the proposed sensor and production technologies are based on thin foils and are designed for small to medium measurement areas with the potential for a high areal resolution. For larger areas, fiber or textile-based solutions can be applied. Textile-based solutions have the advantage of seamless integration in clothing systems or compression garments and high wearing comfort.

9.2 FLEXIBLE SUBSTRATES FOR PRESSURE SENSORS

Conventional pressure sensors are mostly rigid and work with a connection to bulky electronics. For the successful development of flexible sensors and sensor arrays, two main challenges need to be faced: (1) flexible substrates and (2) translation of the relevant physical measurement principles to the substrate.

Advances in manufacturing processes allowed a size reduction of the sensors and the needed electronics. Historically, most flexible sensors for wearable applications are based on foils with printed structures or applying technologies such as photolithography or etching (Xu et al. 2018). More recently substrates such as PDMS (polydimethylsiloxane), PET (polyethylene terephthalate), PI (polyimide), or PEN (polyethylene naphthalate) are used as the base for sensing elements which are then combined with carbon nanotubes, grapheme, or other conductive materials as electrodes. Alternatively, also textile structures can be used as flexible substrates. Sensors can be fiber based or done by textile structures or the assembly of fabrics.

Developing pressure sensors that are flexible and stretchable expands their application on non-flat and rugged surfaces. This is especially important in the field of robotic applications or when used in close contact with the human skin such as for compression garments.

9.3 PHYSICAL PRINCIPLES FOR PRESSURE SENSING

Some of the common types of conventional pressure sensors and their working principle can be transferred to flexible structures including textile or fiber-based solutions. Most sensors for pressure detection are based on changes in electrical resistance, piezo resistance, or electrical capacity. Also, piezoelectrical sensors and other active or more exotic effects can be used, such as the Hall effect, the Gauss effect, the magneto-restive effect, or pyro-electrical and thermo-electrical effects. Also, mechano-chromic effects (e.g. FRET) can be used to assess pressure or strain.

The most common work principles are illustrated in Figure 9.1 (Chen and Yan 2020). In the case of sensors based on resistive effects and an applied constant voltage, the measured current will be related to the mechanical force on the sensing element. Capacitive sensors are based on two electrodes with a dielectric layer in between. Applied external forces will result in changes in the distance between

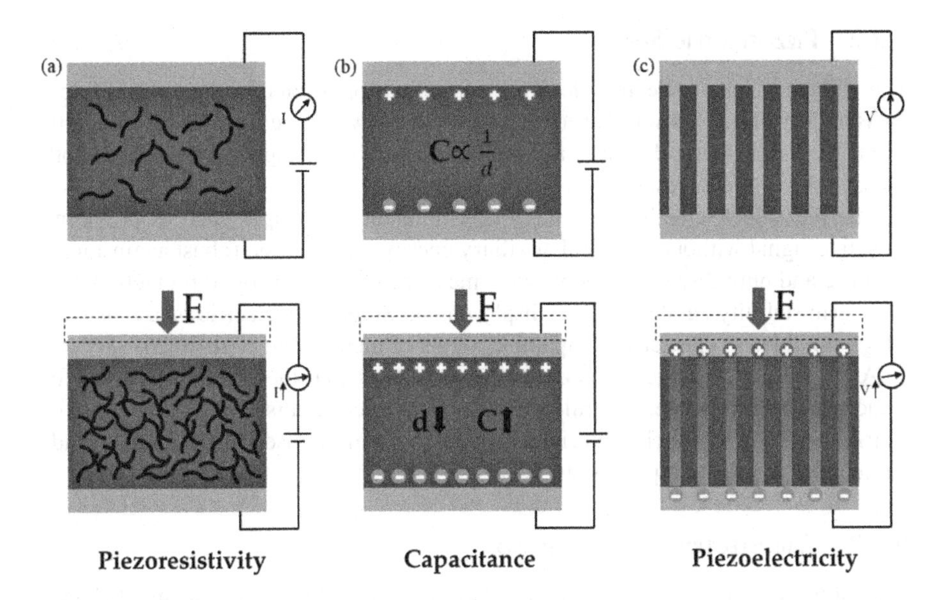

FIGURE 9.1 Working principle of flexible pressure sensors according to Chen and Yan (2020, figure licensed under CC BY 3.0): (a) piezoresistivity, (b) capacitance, (c) piezoelectricity.

the electrodes which will change the capacitance in turn. A pressure applied to the piezoelectric material will induce electrical charge separation, which will result in a measurable change in electrical tension over the sensing element.

9.3.1 Piezoresistive Sensors

Strain applied to conducting and semi-conducting materials will affect the inter-atomic spacing which may change the bandgaps corresponding to the energy difference between valence and conduction bands. This means a higher or lower activation energy is needed for electrons to be raised into the conduction band (depending on the type of material and applied strain) which results in a change in the resistivity of the material. Within the measurement range of the strain sensor, this relationship is linearly related to the piezoresistive coefficient:

$$\rho_\sigma = \frac{\left(\frac{\partial_\rho}{\rho}\right)}{\varepsilon}$$

with

$\partial \rho$ Change in resistivity [Ω ·m]
ρ Original resistivity [Ω ·m]
ε Strain [m/m] or [unit-less]

9.3.2 Piezoelectric Sensors

The voltage at a piezoelectric element is directly proportional to the applied force, pressure, or strain. Sensors based on this technology can be used to assess multiaxial forces (transverse, longitudinal) when the resulting voltage is measured on both axes, or shear effects respectively can be measured.

The piezoelectric technology has many inherent advantages such as providing an active signal without electrical auxiliary energy, ruggedness (robust against temperature and humidity fluctuations, wide measuring range, overload protection, and long-term stability), and almost no displacement. Some piezoelectric materials have a high modulus of elasticity, comparable to that of metals (Wu et al. 2021).

A major disadvantage of piezoelectric sensors is that they cannot be used for static measurements over several hours. A static force results in a fixed amount of charge on the piezoelectric material which will be compromised by the applied readout technology due to input resistors.

9.3.3 Triboelectric Pressure Sensor

The triboelectric effect, or triboelectric charging, relates to the mechanism of charge generation on the surface of certain materials when they are brought into frictional contact with specific other materials. Flexible triboelectric nanogenerators (TENG) can be based on fibers, fabrics, or foils as these configurations have the potential of surfaces rubbing against each other. Pressure applied to such a sensing element results in relative movement of fibers or textile layers, which generates a voltage, signal proportional to the force applied (Zhang et al. 2022).

As triboelectric sensors generate voltage signals in response to a physical effect by charging the contacts, they can work without an external power supply.

9.3.4 Capacitive Sensors

Capacitance is defined as the ratio of the amount of electric charge stored on a conductor to a difference in electric potential (see Figure 9.1b).

$$C = \varepsilon_0 \frac{A}{d}$$

C capacitance (farad)
ε_0 electric constant ($\approx 8.854 \times 10^{-12}$ F/m)
A area of conductive plate (m^2)
d distance between plates (m)

Both the electrically conductive layers and the dielectric layer can be realized using flexible materials or textiles (Grancarić et al. 2018). By varying the geometry (area A and distance d) or the dielectric properties of the used materials the measurement range and sensitivity can be tailored (Cerovic, Petronijevic, and Dojcilovic 2014). Especially parameters related to the solid volume fraction of a fabric such as fineness of the yarn, thread count, and construction will influence the dielectric property (Mukai, Dickey and Suh 2020).

9.3.5 Optical and Multimaterial Fibers

Optical fibers are used for strain or pressure detection either directly in the case of flexible polymer optical fibers (Krehel et al. 2013) and the changed transmission due to pressure-related changes in diameter, or functionalization such as fiber Bragg grating (FBG) (Sahota, Gupta and Dhawan 2020).

One of the main advantages of optical fiber-based sensors is the potential for seamless integration in fabrics and the possibility of manufacturing sensors of deliberate size according to the applied textile manufacturing process. Different types of polymer optical fibers (POF) have been recently developed using materials such as poly(methyl methacrylate) (PMMA), polystyrene (PS), silicon-based elastomers, cyclic polyolefins, or fluorinated polymers.

Classical optical fibers are produced by thermoplastically pulling a preform of heated silica (Yan et al. 2020). The same technique can be used to produce multimaterial fibers with a defined cross-section (Qu et al. 2018). The preform can be produced with different techniques such as thin-film rolling or additive manufacturing and is then heated in a furnace to soften the materials. Pressure sensors can be obtained by arranging polymer composite electrodes within a soft thermoplastic elastomer support (Leber et al. 2020) and subsequent thermal drawing.

9.3.5.1 Change in Diameter

Changing the shape of the cross-section of a soft polymer optical fiber changes the amount of light transferred through the fiber (Wang et al. 2005). Wang et al. demonstrated that an array of optical fibers lying in perpendicular rows and columns separated by elastomeric pads can be used to assess shear forces and plantar pressure in the lower limb.

Krehel et al. (2013), as well as Quandt et al. (2017), showed successfully that fibers out of different elastic and transparent copolymer materials could be extruded and tested for pressure sensing applications (see Figure 9.2). Quandt et al. (2017) produced a demonstrator with a matrix of intersecting fibers secured on a textile

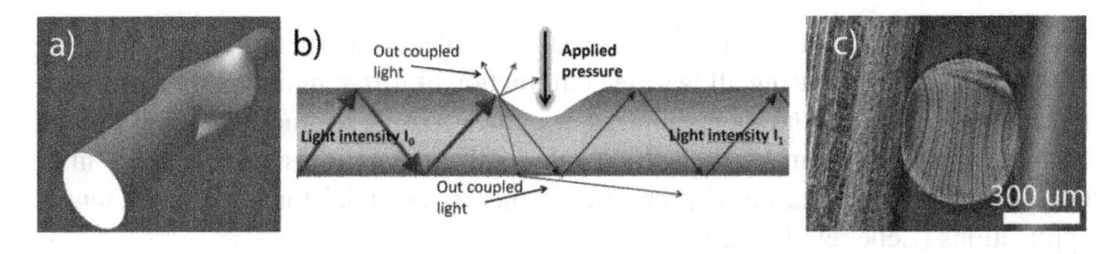

FIGURE 9.2 Pressure sensor based on flexible optical fibers (Krehel et al. 2013, figure licensed under CC BY 3.0): (a,b) influence of applied pressure on elastic optical fiber, (c) deformation of a fiber demonstrated with a micrograph.

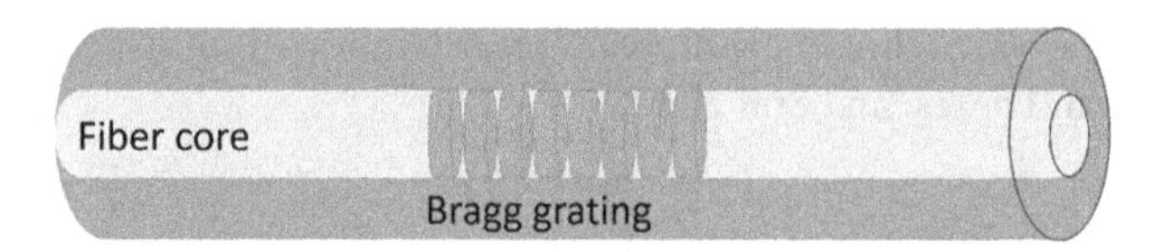

FIGURE 9.3 Fiber Bragg grating (FBG).

substrate which she tested successfully for pressure localization and distribution assessment.

9.3.5.2 Functionalization

Functionalized optical fibers can be used as strain sensors by periodic modulation of the refractive index in specific segments of the fiber. A so-called fiber Bragg grating (FBG) which is a type of distributed Bragg reflector, is constructed in a short segment of the optical fiber reflecting specific wavelengths of light while transmitting all others. This is achieved by periodic variation of the refractive index of the fiber core (see Figure 9.3), which generates a wavelength-specific dielectric mirror (Gao, Liu, et al. 2022).

According to the mechanical strain applied to the fiber, the wavelengths transmitted or reflected respectively will change which allows for determining the force.

9.3.6 Pneumatic Sensors

Pneumatic sensors are mainly used to measure the pressure under a compression garment (Kokai et al. 2021) or between two surfaces under pressure. This type of sensor requires an air-tight structure filled with air and a reference pressure sensor. In the easiest case, pressure detection is done by detecting contact between two surfaces dependent on the applied pressure. The air pressure within the air-filled structure is related to the pressure applied and can be measured with the reference sensor. Despite the simplicity of a single sensor, it is difficult to construct a high-resolution array of sensor elements using this technology as small air channels can be blocked or lead to air resistance which would bias the measurement.

9.4 TEXTILE INTEGRATION OF PRESSURE SENSORS

Textile sensing elements can be fiber, fabric, or garment construction based. State-of-the-art fiber manufacturing allows the combination of different polymers within one fiber or yarn. Apart from melt spinning and fiber coating technologies (e.g. plasma functionalization (Galinski et al. 2020)) thermal drawing is used to produce multimaterial fibers with targeted properties, which is essential for fiber-based sensor applications (Leber et al. 2020).

Possanzini et al. (2019) showed that the measurement range and sensitivity of textile-based pressure sensors can be adjusted by controlled changes to the fiber and fabric properties. Su et al. (2022) (see Figure 9.4) described in their review that the sensitivity and measurement range of their capacitive pressure sensor could be

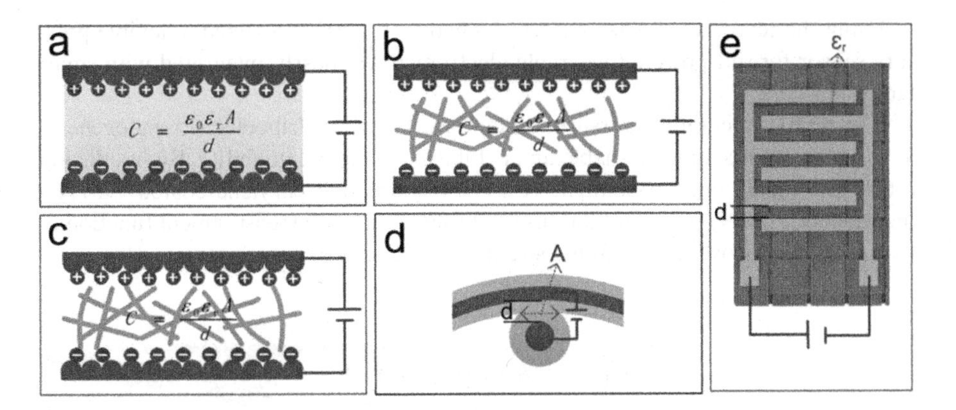

FIGURE 9.4 Capacitive pressure sensors based on textile structures described by Su et al. (2022, figure licensed under CC BY 4.0): (a) textile structured electrodes, (b) textile-structured dielectric layer, (c) combination, (d) yarn structure, (e) in-plane structure.

adjusted by the textile structuring of the electrodes, the dielectric layer, the yarn, and the in-plane structure.

Wang et al. (2016) used carbonized silk fabric (CSF) to produce a textile-based strain sensor. In their work, they demonstrated that the resulting fabrics could be used as ultra-stretchable and highly sensitive strain sensors, allowing the reliable detection of both large and subtle human motions.

Atalay et al. (2018) built a soft sensor based on conductive fabrics with a silicone elastomer-based dielectric layer in-between (see Figure 9.5). To increase the sensitivity they incorporated holes into the dielectric layer.

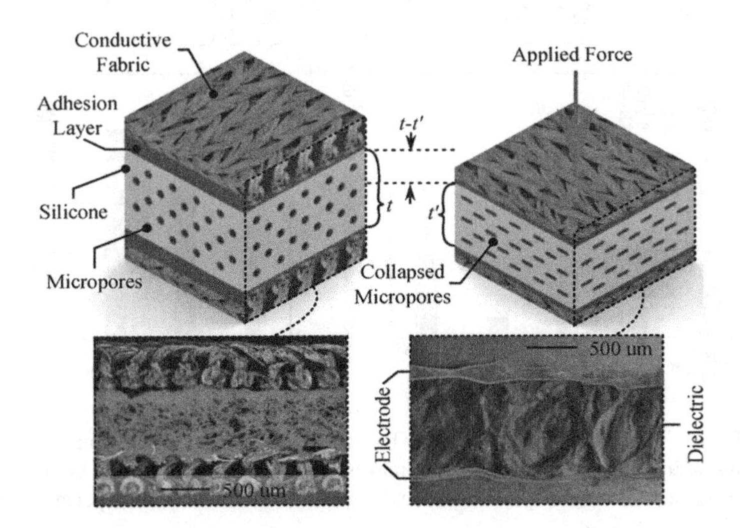

FIGURE 9.5 Soft pressure sensor based on conductive fabrics and dielectric layer. (From Atalay et al. 2018. With permission.)

Using the textile's triboelectric effect Zhao et al. (2020) realized a flexible pressure sensor for self-powered wearable electronics seamlessly integrated with smart fabrics.

The textile-integrated pressure sensors are based on triboelectric nanogenerators, which are machine washable and exhibit excellent breathability. By interlacing robust Cu-coated polyacrylonitrile (Cu-PAN) yarns and parylene-coated Cu-PAN (parylene-Cu-PAN) yarns the authors could show pressure measurement functionality for stitched, woven, and knitted structures (Figure 9.6).

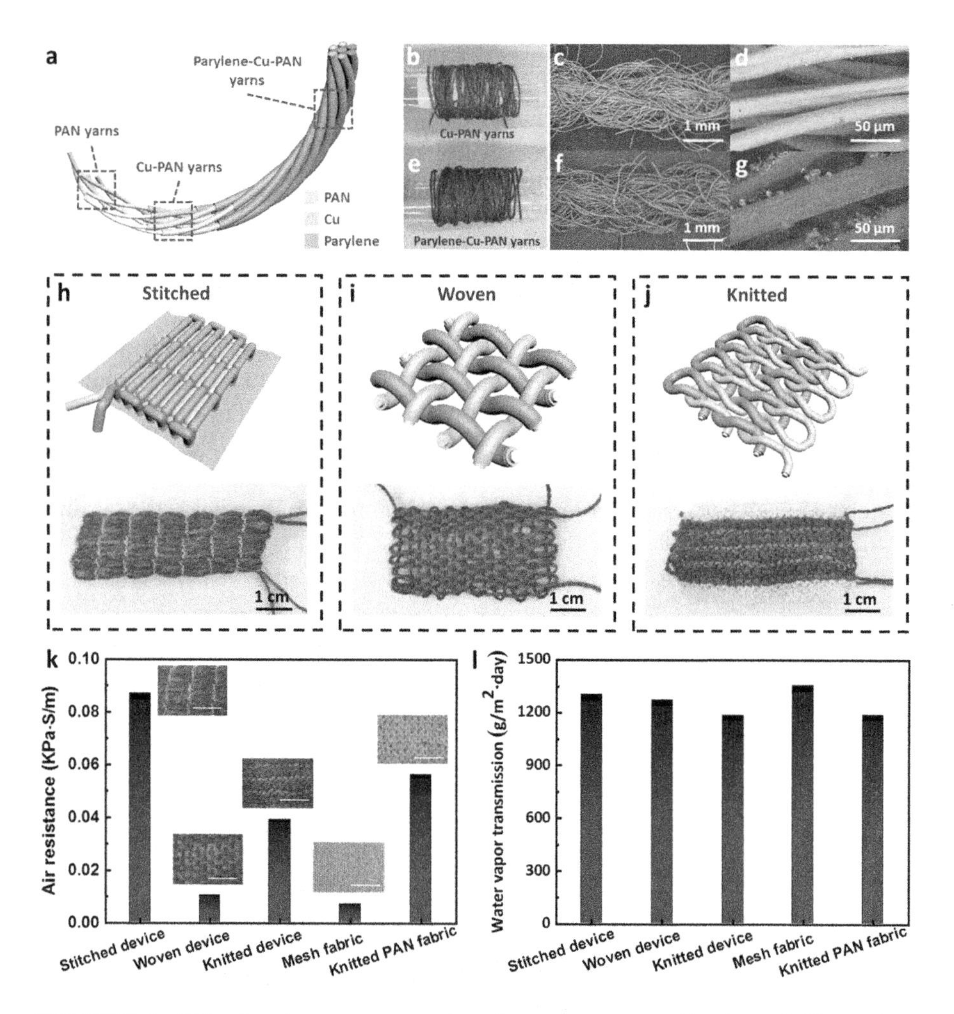

FIGURE 9.6 Triboelectric nanogenerator-based pressure sensors: (a) schematic illustration, (b), (e) digital images, (c), (d), (f), (g) SEM images, (h–j) stitched, woven, and knitted samples, (k) air resistance, (l) water vapor resistance. (From Zhao et al. 2020. With permission.)

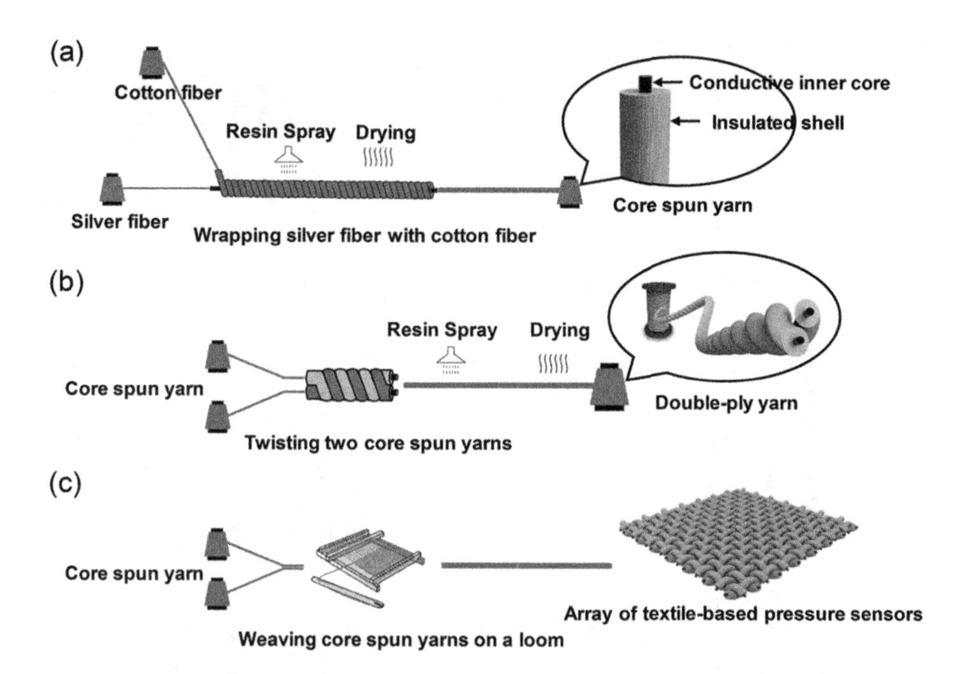

FIGURE 9.7 Fiber-based pressure sensor made from twisted core-spun yarns composed of conductive and isolating elements: (a) core-spun yarn based on silver-coated nylon fiber wrapped with cotton, (b) double-ply yarn sensor: capacitive structure with two core-spun yarns into a double-helix, (c) array of textile-based pressure sensor by weft and warp weaving of two core-spun yarns. (From Zhang et al. 2019. With permission.)

Zhang et al. (2019) developed fiber-based sensing elements based on twisting two core-spun yarns into a fine double-ply yarn (Figure 9.7). They could show that such sensors integrated into gloves or kneepads exhibited reliable measurements without compromising wear comfort or freedom of movement.

9.5 OVERVIEW OF EXISTING/COMMERCIAL SENSORS

The overview of existing, commercially available, flexible pressure sensors was made by internet search and references in scientific papers in the timeframe of Spring to Summer 2022 (Table 9.1). The search was done by looking for flexible pressure sensors, compression garments, and smart textiles and foils for body monitoring or robotic applications. Scientific paper-related links were traced back to the manufacturer if still valid.

So far, only a limited number of real textile or flexible foil-based systems are available commercially (e.g. Sensomative, or Leap Technology). Most commercial products are used in the context of pressure stockings and verification of the applied pressure.

TABLE 9.1
Overview of Commercial Flexible Pressure Sensors

Phys. mechanism	*Name*	*Company*	*Measurement range*	*accuracy*
Pneumatic	Kikuhime®	TT Medittrade/ZiboCare Denmark, Horens, Denmark	0–120 mmHg	±8 mmHg
	PicoPress®	Microlab Electronica, Ponte S.Nicolo PD, Italy	0–189 mmHg	±3 mmHg @10–30°C
	SIGaT-tester® (Partsch and Mosti 2010)	Ganzoni-Sigvaris, St Gallen, Switzerland		
	MST MPF-7®	Swisslastic AG, St Gallen, Switzerland	5–70 mmHg	–
	Talley Digital Skin Evaluator SD500®	Talley Medical Group Ltd, Ramsey, UK		
	Juzo Pressure Monitor®	Compression Innovations, Inc., Cuyahoga Falls, OH, USA [a]	–	–
Piezoelectric	Smart Sleeve Pressure Monitor®	Carolon, Rural Hall, NC, USA[b]		- / 5 mmHg resolution
Optical	Diastron®	Dia-Stron Limited, UK		
	OPP-GF	Opsens Solutions Inc. Québec, Canada [c]	172 kPa	±1 kPa
	Model 60	RJC Enterprises LLC, Bothell, WA, USA [d]	-50–300 mmHg	±2 mmHg
	FOP-M200	FISO Technologies Inc., Québec, Canada [e]	±300 mmHg	±1 mmHg
Resistive/ piezoresistive	I-scan, F-Scan System®	Tek.scan Inc., Boston, MA, USA [f]	From 0–28 kPa to 0–207 MPa	- / 8 bit resolution
	QTC™ SP200-10	Paratech Holdco Ltd. Richmond, UK [g]	2 kg	- / 10g activation
	Half Inch ThruMode FSR	Sensitronics, Inc., Bow; WA, USA [h]	0.3–30 psi	- / ±5 % repeatability
	Tactilus®	Sensor Products Inc., Madison, NJ, USA [i]	0.1–200 psi	±10 %
	Flexiforce® A301	Tekscan Inc., Boston, MA, USA [j]	From 0–4 N to 0–445 N	<8% of the full scale
	FSR Series, FSR03	Ohmite, Warrenville, IL, USA	5 kg	- / 10g activation
	Sensomative Office	Sensomative, Rothenburg, Switzerland	1–100 kPa	- / 8 bit resolution

(*Continued*)

TABLE 9.1 CONTINUED
Overview of Commercial Flexible Pressure Sensors

Phys. mechanism	*Name*	*Company*	*Measurement range*	*accuracy*
Capacitive	Force-sensitive capacitors, FSC16	LEAP Technology, Aabenraa, Denmark [k]	From 0–10 N to 0–1500 N	-
	texsens®-g	Novel GmbH, Munich, Germany [l]	1–10 kPa	–
	pliance® seat	Novel GmbH, Munich, Germany [m]	2–60 kPa	–
	SingleTact, S15-4.5N cal	PPS UK Limited, Glasgow, UK [n]	4.5 N	- / 10 mN activation
	iLoad Flex	Loadstar Sensors, Fremont, CA, USA [o]	10 / 100 / 1000 lb	±5% of the full scale
	x-series	XSensor, Calgary, Canada [p]	0.1–500 psi	±3% of the full scale
Other technologies	QTSS force and shear sensor (Dejke et al. 2021)	Quantum Technology Super Sensors NETpark, Thomas Wright Way, Sedgefield, Co. Durham, TS21 3FD, UK[q]		

[a] https://designinterface.com/medical-devices/
[b] https://www.veindirectory.org/magazine/article/techniques-technology/monitoring-compression-dosages
[c] https://opsens-solutions.com/products/fiber-optic-pressure-sensors/opp-gf/
[d] http://www.rjcenterprises.net/products.html
[e] https://fiso.com/de/service/medizin/
[f] https://www.tekscan.com/products-solutions/pressure-mapping-sensors/9801
[g] https://www.peratech.com/assets/uploads/datasheets/Peratech-QTC-DataSheet-SP200-Series-Nov15.pdf
[h] https://www.sensitronics.com/products-half-inch-thru-mode-fsr.php
[i] https://www.sensorprod.com/freeform.php
[j] https://www.tekscan.com/products-solutions/force-sensors/a301
[k] https://leaptechnology.com/product/force-sensitive-capacitor
[l] https://www.novel.de/products/texsens/
[m] https://www.novel.de/products/pliance/
[n] https://www.singletact.com/micro-force-sensor/calibrated-sensors/15mm-calibrated-sensors/15mm-4-5-newton/
[o] https://www.loadstarsensors.com/assets/iload-flex.pdf
[p] https://www.xsensor.com/solutions-and-platform/sensors
[q] www.quantumtechnologysupersensors.com

9.6 CURRENT RESEARCH AND FUTURE TRENDS

Currently, efforts are made in the field of functional fibers and the development of flexible electronics which can be seen in the high number of scientific papers in recent years (Web of Science search for flexible pressure sensors showed an increase from the first papers in 1984 to 1988 in 2021). Such solutions will potentially be available as commercial products in the coming years, but so far, only a few commercial sensors can be found on the market.

Unobtrusive, fully textile-integrated pressure or strain sensors could be used for patient monitoring, rehabilitation purposes, or performance monitoring in sports applications. Robustness and repeatability (Zhang et al. 2022) including potential cross-talk (Xu et al. 2018) between sensors of an array are still a reported challenge. Su et al. (2022) report potential issues with peeling and unevenness of conductive materials on the surface of the textile-based electrode which will affect the conductivity and durability of the electrodes.

Printed electrical circuits or connections on foils and fabrics are fast-growing, advanced manufacturing technologies (Tan, Tran and Chua 2016). Electrical conductive polymers and adapted printing technologies (3D printing or jet printing) are used for the development of wearable electronic sensors or connecting sensors to the electronics enabling smart, wearable systems.

With the industrialization of these techniques, manufacturing thin electronics at low cost with high throughput will become possible which will allow the production of highly flexible and stretchable sensors and electronics in general (Soin 2022). The balance of measurement range and sensitivity needs further research as proposed by Zhang et al. (2022). In addition, most of the proposed solutions in the literature are still in the research phase and haven't been checked to ensure fabric sensors be washable, non-toxic, and resistant to surface shear, bending, torsion, and other stimuli during wear and use (Zhang et al. 2022).

9.7 CONCLUSION

General purpose, flexible, and easy-to-use pressure distribution assessment systems are hardly found on a commercial level. As the measurement systems need to be adjusted to the specific application resulting in many specialized systems, the interest of industrial suppliers is limited so far. Commercial solutions are available only for specific applications and are rather expensive. Manufacturing processes reported as simple on a lab scale in papers result in high costs when upscaled, especially when multiple steps need to be done including the connection to electronics.

In addition, calibration of such systems is not simple and needs experience. The lack of standardization (Zhang et al. 2022, Kokai et al. 2021) of such new products limits the possibility to sell sensors to be incorporated into larger systems.

As compression garments for sports and medical applications have the potential to become a huge market this will change.

The literature overview showed that with a steadily growing market for wearable systems and body monitoring applications on one side and technological progress regarding the size and weight of the needed electronics on the other side, it becomes easier to produce flexible sensor systems.

De Pasqual (De Pasquale and Ruggeri 2019) stated in his paper that "fabrics are the new silicon". With the possibility to combine different materials in defined, flexible yarns and fabric structures and geometries, sensing elements can be produced at the sub-millimeter level that can also be combined into arrays.

ABBREVIATIONS

DC: direct current
FBG: fiber Bragg grating
FRET: Förster resonance energy transfer
PDMS: polydimethylsiloxane
PEN: polyethylene naphthalate
PET: polyethylene terephthalate
PI: polyimide
PMMA: poly(methyl methacrylate)
POF: polymer optical fibers
SEM: scanning electron microscope
TENG: triboelectric nanogenerators

REFERENCES

Atalay, Ozgur, Asli Atalay, Joshua Gafford, and Conor Walsh. 2018. "A highly sensitive capacitive-based soft pressure sensor based on a conductive fabric and a microporous dielectric layer." *Advanced Materials Technologies* 3 (1):1700237. doi: 10.1002/admt.201700237.

Cerovic, Dragana D, Ivan Petronijevic, and Jablan R Dojcilovic. 2014. "Influence of temperature and fiber structure on the dielectric properties of polypropylene fibrous structures." *Polymers for Advanced Technologies* 25 (3):338–342. doi: 10.1002/pat.3245.

Chen, Wufan, and Xin Yan. 2020. "Progress in achieving high-performance piezoresistive and capacitive flexible pressure sensors: A review." *Journal of Materials Science & Technology* 43:175–188. doi: 10.1016/j.jmst.2019.11.010.

De Pasquale, Giorgio, and Valentina Ruggeri. 2019. "Sensing strategies in wearable biomechanical systems for medicine and sport: A review." *Journal of Micromechanics and Microengineering* 29 (10):103001. doi: 10.1088/1361-6439/ab2f24.

Dejke, Valter, Mattias P. Eng, Klas Brinkfeldt, Josephine Charnley, David Lussey, and Chris Lussey. 2021. "Development of prototype Low-Cost QTSS™ wearable flexible more enviro-friendly pressure, shear, and friction sensors for dynamic prosthetic fit monitoring." *Sensors* 21 (11):3764.

Galinski, Henning, Daniel Leutenegger, Martin Amberg, Fabio Krogh, Volker Schnabel, Manfred Heuberger, Ralph Spolenak, and Dirk Hegemann. 2020. "Functional coatings on high-performance polymer fibers for smart sensing." *Advanced Functional Materials* 30 (14):1910555. doi: 10.1002/adfm.201910555.

Gao, Wenjing, Jianxia Liu, Huiyong Guo, Xin Jiang, Shaofa Sun, and Haihu Yu. 2022. "Multi-Wavelength Ultra-Weak Fiber Bragg Grating Arrays for Long-Distance Quasi-Distributed Sensing." *Photonic Sensors* 12 (2):185–195. doi: 10.1007/s13320-021-0635-4.

Grancarić, A M, I Jerković, V Koncar, C Cochrane, F M Kelly, D Soulat, and X Legrand. 2018. "Conductive polymers for smart textile applications." *Journal of Industrial Textiles* 48 (3):612–642. doi: 10.1177/1528083717699368.

Kokai, Orsolya, Sharon L. Kilbreath, Patrick McLaughlin, and Elizabeth S. Dylke. 2021. "The accuracy and precision of interface pressure measuring devices: A systematic review." *Phlebology* 36 (9):678–694. doi: 10.1177/02683555211008061.

Krehel, Marek, René M. Rossi, Gian-Luca Bona, and Lukas J. Scherer. 2013. "Characterization of flexible copolymer optical fibers for force sensing applications." *Sensors* 13 (9). doi: 10.3390/s130911956.

Leber, Andreas, Alexis Gérald Page, Dong Yan, Yunpeng Qu, Shahrzad Shadman, Pedro Reis, and Fabien Sorin. 2020. "Compressible and electrically conducting fibers for large-area sensing of pressures." *Advanced Functional Materials* 30 (1):1904274. doi: 10.1002/adfm.201904274.

Mukai, Y, E C Dickey, and M Suh. 2020. "Low frequency dielectric properties related to structure of cotton fabrics." *IEEE Transactions on Dielectrics and Electrical Insulation* 27 (1):314–321. doi: 10.1109/TDEI.2019.008511.

Partsch, Hugo, and Giovanni Mosti. 2010. "Comparison of three portable instruments to measure compression pressure." *International Angiology: A Journal of the International Union of Angiology* 29:426–430.

Possanzini, Luca, Marta Tessarolo, Laura Mazzocchetti, Enrico Gianfranco Campari, and Beatrice Fraboni. 2019. "Impact of fabric properties on textile pressure sensors performance." *Sensors* 19 (21):4686.

Qu, Yunpeng, Tung Nguyen-Dang, Alexis Gérald Page, Wei Yan, Tapajyoti Das Gupta, Gelu Marius Rotaru, René M Rossi, Valentine Dominique Favrod, Nicola Bartolomei, and Fabien Sorin. 2018. "Superelastic multimaterial electronic and photonic fibers and devices via thermal drawing." *Advanced Materials* 30 (27):1707251. doi: 10.1002/adma.201707251.

Quandt, B M, R Hufenus, B Weisse, F Braun, M Wolf, A Scheel-Sailer, G-L Bona, R M Rossi, and L F Boesel. 2017. "Optimization of novel melt-extruded polymer optical fibers designed for pressure sensor applications." *European Polymer Journal* 88:44–55. doi: 10.1016/j.eurpolymj.2016.12.032.

Sahota, Jasjot, Neena Gupta, and Divya Dhawan. 2020. "Fiber Bragg grating sensors for monitoring of physical parameters: A comprehensive review." *Optical Engineering* 59 (6):060901.

Soin, Norhayati. 2022. "Recent progress in printed physical sensing electronics for wearable health-monitoring devices: A review." *IEEE Sensors Journal*:1530–437X. doi: 10.1109/JSEN.2022.3142328.

Su, Min, Pei Li, Xueqin Liu, Dapeng Wei, and Jun Yang. 2022. "Textile-based flexible capacitive pressure sensors: A review." *Nanomaterials* 12 (9):1495.

Tan, H W, T Tran, and C K Chua. 2016. "A review of printed passive electronic components through fully additive manufacturing methods." *Virtual and Physical Prototyping* 11 (4):271–288. doi: 10.1080/17452759.2016.1217586.

Wang, Chunya, Xiang Li, Enlai Gao, Muqiang Jian, Kailun Xia, Qi Wang, Zhiping Xu, Tianling Ren, and Yingying Zhang. 2016. "Carbonized silk fabric for ultrastretchable, highly sensitive, and wearable strain sensors." *Advanced Materials* 28 (31):6640–6648. doi: 10.1002/adma.201601572.

Wang, Wei-Chih, William R Ledoux, Bruce J Sangeorzan, and Per G Reinhall. 2005. "A shear and plantar pressure sensor based on fiber-optic bend loss." *Journal of Rehabilitation Research & Development* 42 (3):315–326.

Wu, Yongling, Yulin Ma, Hongyu Zheng, and Seeram Ramakrishna. 2021. "Piezoelectric materials for flexible and wearable electronics: A review." *Materials & Design* 211:110164. doi: 10.1016/j.matdes.2021.110164.

Xu, Fenlan, Xiuyan Li, Yue Shi, Luhai Li, Wei Wang, Liang He, and Ruping Liu. 2018. "Recent developments for flexible pressure sensors: A review." *Micromachines* 9 (11):580.

Yan, Wei, Chaoqun Dong, Yuanzhuo Xiang, Shan Jiang, Andreas Leber, Gabriel Loke, Wenxin Xu, Chong Hou, Shifeng Zhou, Min Chen, Run Hu, Perry Ping Shum, Lei Wei, Xiaoting Jia, Fabien Sorin, Xiaoming Tao, and Guangming Tao. 2020. "Thermally drawn advanced functional fibers: New frontier of flexible electronics." *Materials Today* 35:168–194. doi: 10.1016/j.mattod.2019.11.006.

Zhang, H, L Lin, N Hu, D Yin, W Zhu, S Chen, S Zhu, W Yu, and Y Tian. 2022. "Pillared carbon tungsten decorated reduced graphene oxide film for pressure sensors with ultra-wide operation range in motion monitoring." *Carbon* 189:430–442. doi: 10.1016/j.carbon.2021.12.080.

Zhang, Jia-wen, Yan Zhang, Yuan-yuan Li, and Ping Wang. 2022. "Textile-based flexible pressure sensors: A review." *Polymer Reviews* 62 (1):65–94. doi: 10.1080/15583724.2021.1901737.

Zhang, Qi, Yu Lu Wang, Yun Xia, Peng Fei Zhang, Timothy V Kirk, and Xiao Dong Chen. 2019. "Textile-only capacitive sensors for facile fabric integration without compromise of wearability." *Advanced Materials Technologies* 4 (10):1900485. doi: 10.1002/admt.201900485.

Zhao, Zhizhen, Qiyao Huang, Casey Yan, Youdi Liu, Xiangwen Zeng, Xiaoding Wei, Youfan Hu, and Zijian Zheng. 2020. "Machine-washable and breathable pressure sensors based on triboelectric nanogenerators enabled by textile technologies." *Nano Energy* 70:104528. doi: 10.1016/j.nanoen.2020.104528.

10 Applications in Medical, Sports, and Athletes

Yongrong Wang

CONTENTS

10.1 INTRODUCTION

Compression garments are individually designed and manufactured used to apply substantial mechanical pressure on the surface of needed body regions for stabilizing, compressing, and supporting underlying tissues, including stockings, bandages, sleeves, gloves, bodysuits, and face masks. They have been widely researched and utilized in the fields of medical treatments and rehabilitation (e.g., prevention and treating of chronic venous and lymphatic disorders, scar management, orthopedic support and rehabilitation, and post-operation of plastic surgery), sports and athletics and body-shaping.

The pressure magnitude, distribution, and duration are the most important indicators for therapy efficacy, comfort, and safety, which are determined primarily by the mechanical properties of the fabric and garment style and fit. The pressure should be in an appropriate range. Insufficient pressure will limit its efficiency and perhaps reduce the aesthetic appeal of the wearer, while too much pressure will result in feeling uncomfortable or even breathing difficulty, and thus may lead to serious damage to health. Denton identified the pressure threshold of discomfort to be around 44.1–73.5 mmHg depending on the individual perception and body region, which was greater but close to the average capillary blood pressure of 32.3 mmHg near the skin surface. The pressure comfort range for the normal condition is 14.7–29.4 mmHg, but it also depends on the individual condition (Xiong and Tao 2018, Wang and Gu 2022).

DOI: 10.1201/9781003298526-10

10.2 PREVENTION AND TREATING OF CHRONIC VENOUS AND LYMPHATIC DISORDERS

Chronic venous disease (CVD) is a very common disease, connected to risk factors such as age, sex, family history, obesity, pregnancy, and a standing occupation. Venous disease was divided into C0–C6 grades by CEAP (Clinical-Etiology-Anatomy-Pathophysiology) classification system, which are: C0: No visible or palpable signs of venous disease; C1: Telangiectasis or reticular veins; C2: Varicose veins; C3: Edema; C4: Changes in skin and subcutaneous tissue divided into two subclasses; C4a: Pigmentation or eczema; C4b: Lip dermatosclerosis or atrophies blanche; C5: Healed venous ulcer; C6: Active venous ulcer (Wang and Gu 2022). Compression therapy is the mainstay of precaution or one of the basic treatment methods used for the above venous diseases through the main mechanisms: (1) It reduces the venous diameter and increases the interstitial pressure in the surroundings. This results in increasing blood flow in the deep veins, reducing the amount of pathologic reflux, and decreasing the hydrostatic pressure; (2) It restores the valve function by bringing the walls of the veins closer together; (3) It reduces blood pressure in the superficial venous system; (4) It reduces the pressure differences between the capillaries and the tissue to prevent backflow; (5) It increases the cutaneous microcirculation, favors white cell detachment from the endothelium, and prevents further adhesion; and (6) It reintegrates the interstitial liquids into the vessels (O'Meara et al. 2012).

Lymphedema occurs if the lymphatic system is damaged. Lymph vessels form part of your circulatory system which helps fluid to move around your body. If fluid is unable to effectively drain, lymph fluid builds up in the spaces of the body tissue which causes swelling. If left untreated the swelling will get worse and the condition becomes chronic. Lymphedema mainly occurs in the arms or legs. The four clinical states of lymphedema are listed in Table 10.1, reported in the 《Consensus Document》 of the International Society of Lymphology (Lee and Rockson 2011). Three stage images of lymphedema are shown in Figure 10.1: (a) shows that the edema of human legs completely subsides after positive treatment; (b) shows that the edema of legs partially subsides after negative treatment, and there is still a state of relaxation and swelling; and (c) shows that the overall edema of legs is swollen, like the skin of an elephant.

TABLE 10.1
Staging of the International Society of Lymphology (Lee and Rockson 2011)

Stage 0	Stage 1	Stage 2	Stage 3
Subclinical with possible clinical evolution	Edema regressing with treatments with positive pitting test	Edema partially regressing with treatments with negative pitting test	Elephantiasis with cutaneous complications and recurrent infections

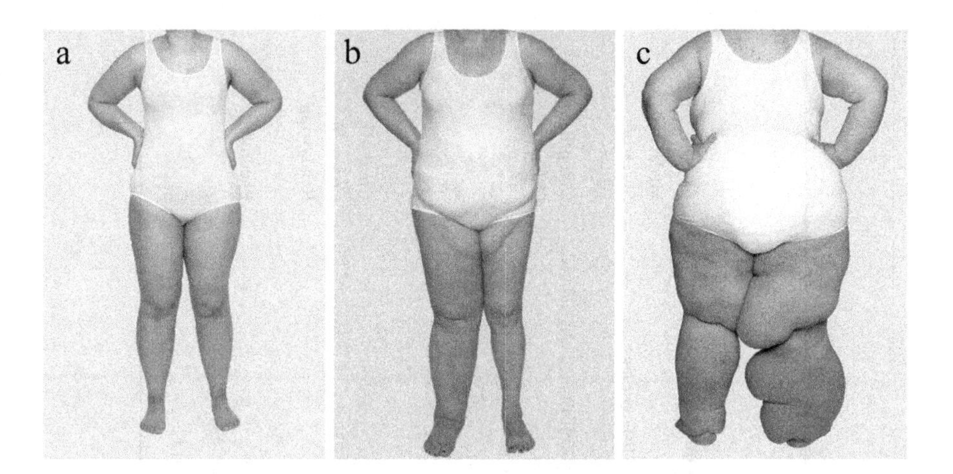

FIGURE 10.1 Staging of the International Society of Lymphology. (a) Stage 1; (b) Stage 2; (c) Stage 3.

Compression therapy is one of the most important components in decongestive lymphatic therapy (DLT), which is the universally accepted basis of the conservative treatment of lymphedema. Compression therapy aims to alleviate symptoms, prevent progression, and reduce the risk of skin infection through four main mechanisms (Lee and Rockson 2011): (1) Compression reduces microcirculatory filtration, due to a shift in Starling's equilibrium. Increasing the tissue pressure will reduce the transmural pressure gradient, resulting in a reduction of the lymphatic load; (2) Compression promotes lymphatic drainage, both in the initial lymphatics and in the lymph collectors, resulting in an augmentation of the lymphatic pump; (3) Compression shifts tissue fluid toward the non-compressed parts of the limb; and (4) Compression softens fibrotic tissue changes.

Compression therapy is divided into two major categories: elastic and inelastic compression therapy (Partsch and Mortimer 2015). In elastic compression therapy, modalities composed of elastic fibers provide compression during movement and at rest. During ambulation, calf muscles contract, accordingly, the modality expands, dissipating the force exerted by muscle contraction. This release in pressure favors for venous return to the heart, such as medical compression stockings, bandages, etc. Conversely, inelastic compression therapy increases pressure and assists drainage and venous support. It produces high pressure with muscle contraction and low pressure at rest. The advantages of inelastic compression therapy include protection against trauma and little interference with daily activities. Its disadvantages include the increased pressure exerted on the leg for an extended period of time and inadequacy of this device for highly exudative wounds. Such as Unna boot, Velcro compression strips, and Intermittent pneumatic compression (IPC).

The advantages and disadvantages of different modalities applied in compression therapy are listed in Table 10.2. The different styles of medical compression stockings, sleeves, Velcro compression straps, and Intermittent pneumatic compressions are shown in Figure 10.2. The adoptions of modalities to a great extent depend upon

TABLE 10.2
Advantages and Disadvantages of Different Modalities

Modalities	Advantages	Disadvantages	Examples
Graduated compression stockings	• No trained physicians required • Suitable for low pressure (20–40 mmHg)	• Difficult to put on • Different stockings for different legs • Pressure loss	Medi® Sigvaris® Jobst®
Compression bandages	• Maintains compression • Pressure can be adjusted • Recommended for a high level of compression (35–80 mmHg)	• Needs to be applied by well-trained physicians and nurses • Pressure variation and no measurement	Comprilan® Coban®
Unna boot	• Inelastic • High working pressure, well tolerated during rest	• Messy • Disposable • Pressure loss	Gelocast®
Velcro compression strips	• Self-application • The pressure loss can be self-adjusted • Build-in pressure system is able to provide information on the local pressure	• Not appealing	CircAid® Compreflex® Readywrap®
Intermittent pneumatic compression (IPC)	• Augments venous return • Effective for immobile patients	• Expensive, noisy, bulky • Requires immobility for a few hours/day	PneumPress®

References: Liu et al. 2017, Rabe et al. 2021, Tiwary 2022, Paranhos et al. 2021, Stather et al. 2019, Nelson et al. 2014.

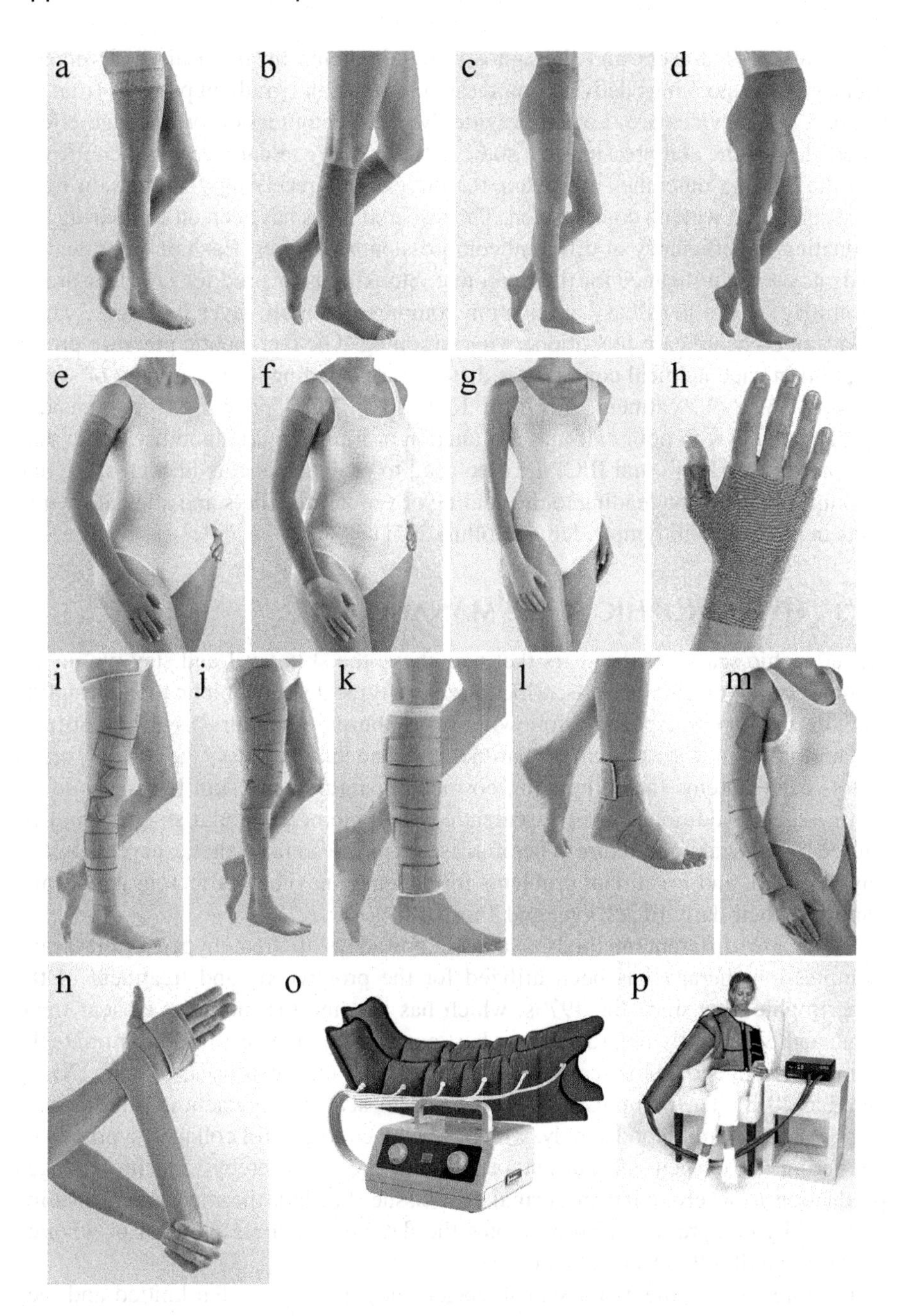

FIGURE 10.2 The different styles of medical compression stockings, sleeves, Velcro compression straps, and intermittent pneumatic compressions. (a) Knee-high with open toe; (b) thigh-high with closed toe; (c) pantyhose with closed toe; (d) maternity panty with closed toe; (e) arm sleeve without hand; (f) arm sleeve with hand; (g) compression glove; (h) hand piece; (i) whole leg; (j) upper leg with knee piece; (k) lower leg; (l) ankle foot wrap; (m) arm sleeve; (n) hand wrap; (o) pump for lower limb; (p) pump for upper limb.

symptoms of end-users and theragnostic requirements by medical professionals. Bandages and stockings deliver compression in static and gradient patterns from toe to knee. IPC devices are used to provide dynamic compression or massage effect; others deliver the compression in a static pattern. Many researchers have confirmed that the healing outcomes are better for the patients receiving compression treatment compared with no compression. The research focus has been on comparing and evaluating the efficiency of different compression modalities. Each of these devices holds its own significance for the treatment. Stocking is favored for low pressure (< 50 mmHg) and allows easy application compared to multi-layer bandages, which need trained healthcare practitioners for wrapping. The therapeutic pressure profile required in each medical condition is different, responding to the severity of symptoms and type of treatment, shown in Table 10.3. IPC is primarily recommended for the patients with poor calf muscle function or limited ankle mobility. More than two modes (stockings and IPC) are also used to provide benefits of both static and dynamic compression leading to the healing of venous disorders and alleviate symptoms in patients with lymphedema (Kolluri 2011).

10.3 HYPERTROPHIC SCARS MANAGEMENT

Hypertrophic scars are the scars that rise above the skin level and stay within the confines of the original lesions, which are very typical skin complications resulting from dermal injury especially following severe burns. The scars are different from the normal skin with thickness, hardness, pigmentation, color, etc., which result in lots of symptoms including pain, cosmetic disfigurement, skin hypersensitivity, itchiness, etc. Additionally, if the scars are close to joints, articular stiffness may be caused by the scar contracture. Therefore, scars may lead to aesthetic, psychological, physiological, and functional problems for patients, inevitably affecting the normal quality of their daily life (Xiong and Tao 2018).

There are different methods available nowadays to treat hypertrophic scars. Compression therapy has been utilized for the prophylaxis and treatment of the hypertrophic scar since the 1970s, which has become the mainstay of scar treatment, and is generally regarded as the best non-invasive means to help minimize the formation of hypertrophic scars and enhance the maturation process of scars (Xiong and Tao 2018). The compression mechanism for scar management is summarized. First, it acts to limit blood supply, which is believed to control collagen synthesis by restricting the oxygen and nutrients carried to the site, thereby reducing collagen production to levels found in normal scar tissue. Second, the mechanical loading imposed by compression also promotes the flattening and realignment of whorled collagen bundles (Crofton et al. 2020).

Compression garments for scar management generally are flat-knitted and then tailor-made into a smaller size than the actual anthropometric measurement of the respective body part in order to deliver pressure onto hypertrophic scars. Various types of garments for different body parts are shown in Figure 10.3, including a mask covering the head and neck and exposing the five features, gloves showing fingertips, exposing the front and back of the limbs and head jumpsuit, tight leggings

TABLE 10.3
Compression Classes and Pressure Profile of MCSs

		Pressure at ankle (mmHg)			Residual pressure ratio of pressure at ankle (%)		
		UK	DE	EU	UK		DE、EU、CN
Class	Compression intensity	B	B	B	C	F	
Ccl I	Low	14–17	18–21	15–21	50–85	50–85*	
Ccl II	Moderate	18–24	23–32	23–32			
Ccl III	High	25–35	34–46	34–46			
Ccl IV	Very high	N/A	>49	>49			

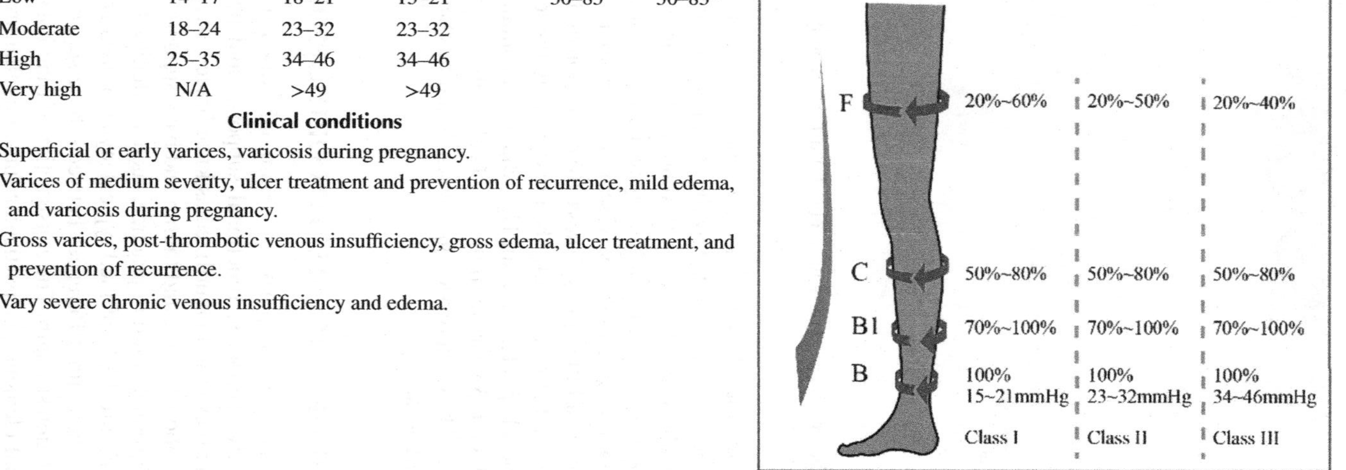

Clinical conditions

Class	Clinical conditions
Ccl I	Superficial or early varices, varicosis during pregnancy.
Ccl II	Varices of medium severity, ulcer treatment and prevention of recurrence, mild edema, and varicosis during pregnancy.
Ccl III	Gross varices, post-thrombotic venous insufficiency, gross edema, ulcer treatment, and prevention of recurrence.
Ccl IV	Vary severe chronic venous insufficiency and edema.

References: Wang and Gu 2022a, Kankariya 2022, Wang and Gu 2022b

UK: British [BS6612:1985;BS661210:2018]; DE: Germany [RAL-GZ387/1: 2008]; EU: Europe [ENV12718:2001]; CN:[YY/T 0853: 2011]. B – ankle with minimum circumference; B1 – the level at which the Achilles tendon changes into the calf muscles; C – the calf region at the maximum circumference; F – the level at mid-thigh.

*Proportion of calf pressure at the thigh (%).

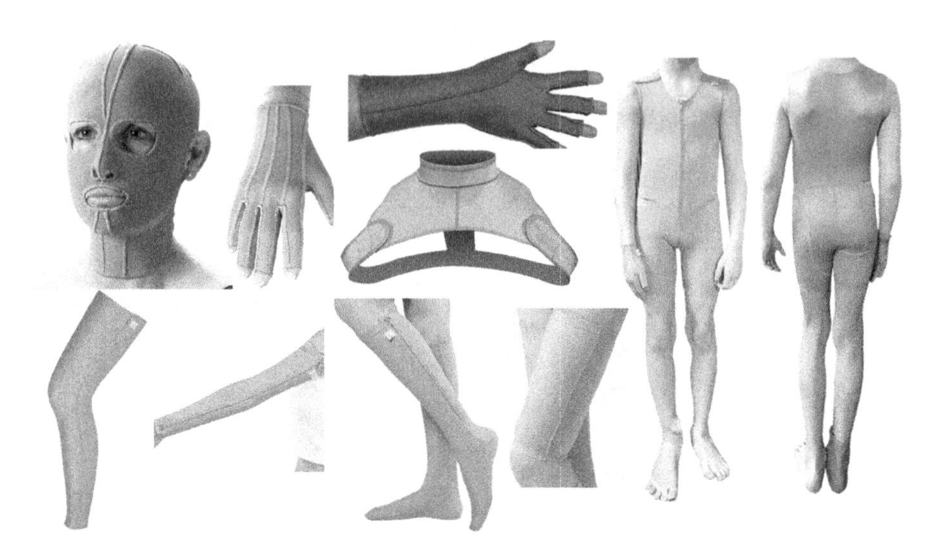

FIGURE 10.3 The pressure therapy garments for different body parts.

and sleeves covering the whole leg and arm; tight calf socks; and tight thigh sleeves. The uniform surface pressure applied by the compression garments can have various positive effects on the scarred area: (1) prevention of uncontrolled growth of the scar; (2) reducing thickness, fading and softening of the scar; (3) relieving pain and itch in the scar area; (4) increasing mobility of scars in joint areas; (5) minimizing risk of scar shrinkage and resulting joint stiffness; (6) protection of the sensitive skin or skin graft. To achieve ideal clinical result, the pressure magnitude delivered by compression garment is best kept at 20–30 mmHg, which is recommended to be worn about 23 h per day and for 1–3 years until the scar heals well without any restrictions with regard to function. The compression garments should be changed every 2–3 months to prevent pressure loss.

10.4 COMPRESSION TEXTILES FOR ORTHOPEDIC SUPPORT

Textile-based orthopedic supports are commonly divided into three categories, including preventive supports, functional supports, and post-operative/rehabilitative supports, and they could be utilized as knee braces, wrist braces, ankle braces, shoulder braces, elbow braces, and thumb and back supports (Xiong and Tao 2018). There are many products used in practice, such as Medi®, Bauerfeind®, McDavid®, LP®, and Cep®. Figure 10.4 outlined typical styles of available orthopedic supports in commercials. The main differences between these supports are the compression size and consolidation strength. The construction of functional compression supports consists of crucial elements for particular functions that are substantial for patient health or the healing process. It should not disturb blood circulation due to pressure to the limb, also it should not be too loose, and not to slip away, and it should not

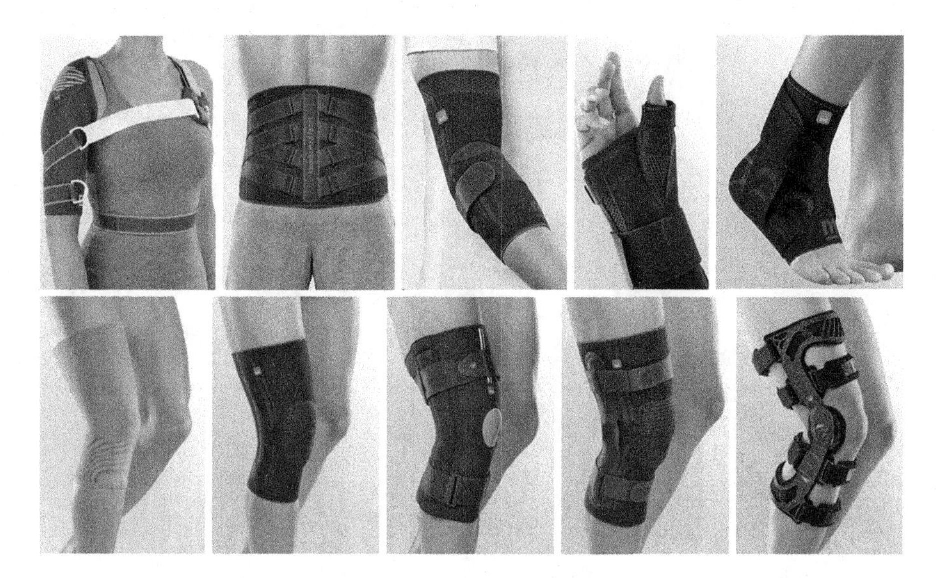

FIGURE 10.4 The typical styles of available orthopedic supports in commercials.

have sharp corners. To date, there is no universally accepted standard in the world in compression requirements and work mechanisms for orthopedic supports to date.

Textile-based orthopedic supports are generally made from an anatomically shaped knitted fabric with other components, such as silicone or other parts for functional application, straps and other fasteners (such as Velcro or similar hook-and-loop types) for fastening the support to the body, which create a positive anatomical change in the body as well as an increase in body strength, enhance motor skills, and/or provide support to paraplegic patients; neonates; elderly, pregnant, and nursing women; and patients with motor disabilities. All parts included support that can change the elasticity of the entire product. Therefore, compression of the orthopedic supports must be evaluated for the final structure of the support, with all additional textile and non-textile parts (Mikucioniene and Muraliene 2022).

10.5 COMPRESSION THERAPY AFTER PLASTIC SURGERY

Compression garments also play a role in cosmetic surgery and are also used for postoperative recovery, for example a compressive bra after a breast augmentation, reduction, lifting, or reconstruction. It offers support to the connective tissue and helps maintain the shape established during the surgery. There is research about the scientific mechanism for the application; however, it is an essential procedure after a tummy tuck or other body-sculpting surgery, which may offer many benefits: (1) The compression garment protects the affected area and allows it to heal, and helps to shorten recovery time. (2) Compression garment helps increase blood circulation to certain parts of the body and reduces the risk of inflammation, swelling, and pain.

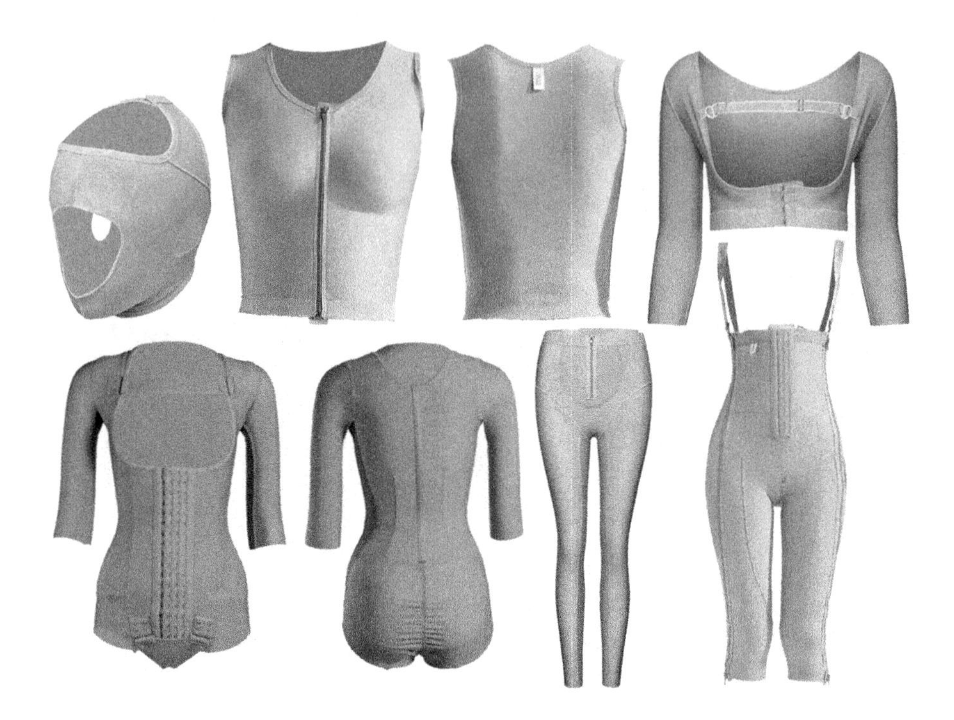

FIGURE 10.5 Compression garments after plastic surgery.

(3) The compression garment helps maintain the new shape you want to achieve through proper support (Kanter plastic surgery Accessed on 10 Dec 2022).

For surgery recovery, there are many available compression garments with different styles generally tailor-made by warp elastic fabric, support bones, or Velcro strip for local pressure strength, hook-and-eye closures or zippers for an adjustable fit, and multiple coverage options with open or accessible crotches, shown in Figure 10.5.

10.6 COMPRESSION SPORTSWEAR

The popularity of the use of compression garments (e.g., whole-body compression clothes, shorts, pants, and socks) during various exercise activities has been increased. Functional compression sportswear is supposed to potentially enhance exercise performance and speed up recovery when worn after endurance or strength exercise. To date, the ideal compression required to be beneficial for performance and recovery has not been standardized. Different conclusions on the effects of wearing compression sportswear may be related to the types of activities and individual differences (Xiong and Tao 2018, MacRae et al. 2011, Fu et al. 2012). The typical compression sportswear in commercials is shown in Figure 10.6, including sport compression socks, calf support, swimwear, compression shorts, and compression pant.

The majority of commercially branded garments currently available for sport applications are claimed to provide the wearer with the promotion of blood flow,

FIGURE 10.6 The typical compression sportswear in commercial.

regulation of blood lactate and creatine kinase levels, better muscle oxygenation, reduced fatigue, improvement of fatigue recovery, reduced muscle oscillation and reduced muscle injury, and the enhancement of muscle functions (e.g., reducing muscle vibrations and activation). Aside from the aforementioned advantages, compression sportswear has also been commonly used and has shown potential benefits in racing sports including speed runs, cycling, skating, and skiing. These broader applications can be attributed to the integrated aerodynamics and mechanical properties of the latest compression sportswear (MacRae et al. 2011, Fu et al. 2012, Fu et al. 2013, Yang et al. 2020).

Some research studies have been conducted to identify the benefits above; another viewpoint is that wearing compression sportswear is not helpful for enhancing sprint or throwing a performance, but could be beneficial to reduce post-exercise trauma, swelling, and perceived muscle soreness, and accelerate the recovery of force production.

Therefore, the design principles of individualized compression sportswear for each discipline have been set according to the following: (1) Applying compression on specific muscles to increase physiological and biomechanical characteristics; (2) Applying the peculiarity of aerodynamics to reduce drag in high-speed sports. Depending on the requirements, the sportswear could be designed according to both principles or individually. In addition to functional features, aesthetics is also an important design criterion in compression sportswear (Xiong and Tao 2018, Fu et al. 2012, Fu et al. 2013).

In addition, for sportswear, the properties of chemical stability, UV resistance, air permeability, water vapor transmission, and washability should also be taken into consideration.

10.7 CONCLUSION

In this chapter, a comprehensive review of the applications of compression garments for medical therapy and sportswear is introduced, including the action principle and modalities for different conditions and body parts, such as the areas of compression garments utilized in chronic venous and lymphatic disorders, scar management, orthopedic supports, after plastic surgery, and sportswear. For the better-wearing instruction and efficiency of compression garments, the pressure criteria for specific modalities and body parts, real-time pressure display, and overpressure warning are the research direction.

REFERENCES

Crofton, E, Meredith, P, Gray, P et al. 2020. "Non-adherence with compression garment wear in adult burns patients: A systematic review and meta-ethnography." *Burns* 46 (2):472–482.

Fu, W, Liu, Y, Zhang, S et al. 2012. "Effects of local elastic compression on muscle strength, electromyographic, and mechanomyographic responses in the lower extremity." *Journal of Electromyography and Kinesiology* 22 (1):44–50.

Fu, W, Liu, Y, Fang, Y et al. 2013. "Research advancements in humanoid compression garments in sports." *International Journal of Advanced Robotic Systems* 10 (1):6.

Kankariya, N. 2022. "Material, structure, and design of textile-based compression devices for managing chronic edema." *Journal of Industrial Textiles* 52 (1):35.

Kanter Plastic Surgery. Accessed on 10 Dec 2022. https://www.kanterplasticsurgery.com/news/compression-garments-after-surgery/.

Kolluri, R. 2011. "Compression therapy for treatment of venous disease and limb swelling." *Current Treatment Options in Cardiovascular Medicine* 13 (2):169–178.

Lee, B B, and Rockson, S G. 2011, 2018. *Lymphedema.* Cham, Switzerland: Spring International Publishing AG.

Liu, R, Guo, X, and Little, T. 2017. "A critical review on compression textiles for compression therapy: Textile-based compression interventions for chronic venous insufficiency." *Textile Research Journal* 87 (9):1121–1141.

MacRae, B A, Cotter, J D, and Laing, R M. 2011. "Compression garments and exercise garment considerations, physiology and performance." *Sports Medicine* 41 (10):815–843.

Mikucioniene, D, and Muraliene, L. 2022. "Influence of orthopedic support structure and construction on compression and behavior during stress relaxation." *Journal of Industrial Textiles* 51 (3_Suppl):5026S–5041S.

Nelson, E A, Hillman, A, and Thomas, K. 2014. "Intermittent pneumatic compression for treating venous leg ulcers." *Cochrane Database of Systematic Reviews* 12 (5):CD001899.

O'Meara, S, Cullum, N, Nelson, E A, and Dumville, J C. 2012. "Compression for venous leg ulcers." *Cochrane Database of Systematic Reviews* 2012 (11). Art. No.: CD000265. doi: 10.1002/14651858.CD000265.pub3.

Paranhos, T, Paiva, C S B, Cardoso, F et al. 2021. "Systematic review and meta-analysis of the efficacy of Unna boot in the treatment of venous leg ulcers." *Wound Repair and Regeneration* 29 (3):443–451.

Partsch, H, and Mortimer, P. 2015. "Compression for leg wounds." *British Journal of Dermatology* 173 (2):359–369.

Rabe, E, Foldi, E, Gerlach, H et al. (2021). "Medical compression therapy of the extremities with medical compression stockings (MCS), phlebological compression bandages (PCB), and medical adaptive compression systems (MAC) S2k guideline of the German Phlebology Society (DGP) in cooperation with the following professional associations: DDG, DGA, DGG, GDL, DGL, BVP." *Hautarzt* 72 (Suppl 2):37–50.

Stather, P W, Petty, C, and Howard, A Q. 2019. "Review of adjustable velcro wrap devices for venous ulceration." *International Wound Journal* 16 (4):903–908.

Tiwary, S K. 2022. *Approach to lower limb oedema.* Singapore: Pringer Nature Singapore Pte Ltd.

Wang, Y R, and Gu, L L. 2022a. "Patient-specific medical compression stockings (MCSs) development based on mathematic model and non-contact 3D body scanning." *The Journal of the Textile Institute.* doi: 10.1080/00405000.2022.2111644.

Wang, Y, and Gu, L L. 2022b. "Predictability of pressure characterization of medical compression stockings (MCSs) directly based on knitting parameters." *Fibers and Polymers* 23 (2):527–536.

Xiong, Y, and Tao, X. 2018. "Compression garments for medical therapy and sports." *Polymers* 10 (6):663.

Yang, C, Xu, Y, Yang, Y et al. 2020. "Effectiveness of using compression garments in winter racing sports: A narrative review." *Frontiers in Physiology* 11 (970):1–10.

11 Efficacy of Compression Garments in Medical and Sports Applications

Kristina Brubacher

CONTENTS

DOI: 10.1201/9781003298526-11

11.1 INTRODUCTION

Compression garments (CGs) have been utilized in the medical sector for many decades, particularly in the treatment of venous dysfunctions (Ramelet 2002, Rohan et al. 2013). Two decades ago, sports scientists discovered CGs as a relatively simple mechanical modality to be applied in sports (Perrey 2008), leading to the development of highly elastomeric sports compression garments (SCGs) designed to compress underlying tissues in an attempt to transfer the beneficial effects of CGs to the sporting environment. A wealth of research exists on CGs for medical and sports applications, but findings are often mixed, and it is not clear how much scientific evidence there is to prove the efficacy of CGs in medical and sports applications. This chapter provides a concise summary of evidence found in existing studies evaluating the functionality of CGs in the treatment of venous disorders and burn scars and in enhancing exercise performance and recovery.

11.2 EFFECTIVENESS OF COMPRESSION GARMENTS AS MEDICAL THERAPY

Traditionally, compression was applied by bandages, meaning its effectiveness was highly dependent on the technique and skills of the bandager (Moffatt 2008). Today, compression therapy is largely provided by medical compression stockings (MCS) in the treatment of venous disorders and thrombosis prevention or custom-made medical compression garments (MCGs) for the treatment of burn scars, which can take various shapes depending on the scarred body part (Macintyre 2007).

Pressure levels applied by CGs are the dosage of compression therapy, which depends on the condition, severity of symptoms, and body part to be treated and is divided into different classes of compression intensity (Ramelet 2002). There are no global standards, leading to variations in norms for compression therapy internationally (Liu et al. 2017). Inadequate pressure levels can adversely affect patients by either not sufficiently treating the condition or causing discomfort or health concerns, such as limb numbness or even cardiac failure through an overload of venous return (Choucair and Phillips 1998, Xiong and Tao 2018).

11.2.1 Evidence for the Effects of Compression Garments in Treating and Preventing Venous Disease

Graduated compression from the ankle (100%) to the thigh (40%) applied by bandages or MCS has been the gold standard therapeutic modality for the treatment, prophylaxis, and management of acute and chronic venous dysfunctions since the 1950s (Ramelet 2002, Rohan et al. 2013). Graduated compression can be applied by elastic or inelastic compression textiles and supports the blood flow up the leg (Partsch 2012). Studies have used a vast array of techniques to investigate the effects of compression on venous and lymphatic function including echotherapy, duplex, and phlebography (Ramelet 2002).

11.2.1.1 Deep Vein Thrombosis

Acute deep vein thrombosis (DVT) is commonly treated with a combination of an anticoagulant and compression therapy to enable patients to be mobile (Galanaud, Laroche and Righini 2013). Wearing MCS in the acute stage of DVT has been shown to reduce pain and swelling (Blättler and Partsch 2003); however, it is currently unclear what the optimal compression treatment involves in terms of stocking length (knee- or thigh-length), pressure level, and application time (Rabe et al. 2018).

MCS are also widely used to prevent DVT in immobile patients, particularly after surgery or trauma (Sajid et al. 2012, Wade, Paton and Woolacott 2017). While some researchers reported that MCS are effective in reducing the risk of DVT post orthopedic surgery (Wells, Lensing and Hirsh 1994, Agu, Hamilton and Baker 1999), others found no effect on preventing DVT after stroke (Dennis et al. 2009) or in elderly patients (Labarere et al. 2006). The varied results could be associated with small sample sizes (Feist, Andrade and Nass 2011) or the heterogeneity of studies. This could include variations in pressure levels depending on the MCS applied in the studies. Best et al. (2000) reported that the pressure delivery of commercial MCS is often inconsistent, which could affect the efficacy.

A review study (Sajid et al. 2012) could not establish the optimal length of MCS for DVT prevention, but Wade et al. (2017) reported a tendency toward knee-high MCS based on patient adherence and preference. New methods of preventing DVT have evolved in recent years, and the efficacy of mechanical compression as DVT prophylaxis has been questioned, particularly when used in addition to anticoagulants (Feist, Andrade and Nass 2011, Rabe et al. 2018).

11.2.1.2 Venous Leg Ulcers

Venous leg ulcers are one of the most common causes of chronic wounds in Europe (Dissemond et al. 2016). Compression plays a key role in the treatment of venous leg ulcers; it is believed to improve patients' quality of life by supporting the healing process, easing pain, and preventing reoccurrences (Dissemond et al. 2016). The use of short-stretch (i.e., inelastic) bandages in multiple layers to create high compression have been recommended (O'Meara et al. 2012). However, MCS have been shown to be as effective or even more effective in supporting venous leg ulcer healing (Junger, Ramelet and Zuccarelli 2004, Amsler, Willenberg and Blättler 2009, Ashby et al. 2014).

It is not clear what compression type provides the most optimal results (Rabe et al. 2018). Many researchers have reported effects of wearing MCS, such as a reduction in perforated veins during daily activities when applying a pressure gradient of 20–30 mmHg (Buhs, Bendick and Glover 1999), hemodynamic improvements (Ibegbuna et al. 2003), and increased arterial perfusion when applying pressure of 13–23 mmHg to forearms (Bochmann et al. 2005). Some researchers believe that these effects are caused by a reduction in transmural pressure and narrowing of veins (Buhs, Bendick and Glover 1999, Bochmann et al. 2005, Rohan et al. 2013). However, a finite-element model showed that deep vein diameter reduction was mainly caused by the contracting muscle rather than MCS (Rohan et al. 2014). More research is needed to establish the underlying mechanism of the effects of compression on the treatment of venous leg ulcers.

11.2.1.3 Edema

Chronic edema is an incurable condition caused by lymphatic failure (lymphedema), vascular insufficiency (e.g., varicose veins), or immobility (dependency edema) (Elwell 2016). Compression therapy is considered the cornerstone treatment in managing venous edema and lymphedema (Mosti and Cavezzi 2019). The treatment of lymphedema consists of two phases: intensive decongestion (short term) and maintenance (long term) and includes other modalities, such as lymphatic drainage (massage), skin care, and exercise (Lee et al. 2013). Compression is critical for both phases and is traditionally applied as multi-layer bandages in the intensive phase applying higher pressure followed by CGs in the maintenance phase to prevent lymph fluid from reaccumulating (Rabe et al. 2018). However, evidence for the most optimal material, pressure level, and application regime is currently lacking. Many studies have shortcomings in their reporting of the details of the methodology, CGs, and pressure application (Mosti and Cavezzi 2019, Kankariya, Laing and Wilson 2021). More randomized controlled trials are needed to establish the optimal conditions for edema management, including considerations of patient mobility, ease of application, and quality of life (Rabe et al. 2018).

11.2.2 Evidence for the Effects of Compression Garments in Treating Burn Scars

Burn injuries causing second-degree or deeper burns commonly result in hypertrophic scars, which substantially reduce skin pliability (Kim et al. 2015). Compression treatment has been the basis of conservative burn treatment to prevent hypertrophic scars since the early 1970s (Puzey 2001, Wienert 2003). Researchers agree that early application of compression until scar maturation is critical for optimal results (Williams, Knapp and Wallen 1998, Puzey 2001). Compression is usually applied by MCGs that are custom-made for the affected body part and need to be worn 23 hours a day for a period ranging from four months to three years (Williams, Knapp and Wallen 1998). The success of compression treatment depends on the age of the patient, ethnicity, location of the burns, and associated mobility (Wienert 2003).

Even though there is clinical and anecdotal evidence to support the use of MCGs in the treatment of burn scars (e.g., Berman and Flores 1998, Yan, Ping and Ping 2010), their efficacy has never been scientifically proven, and the exact mechanism for the treatment of burn scars is yet to be established (Macintyre and Baird 2006, Anzarut et al. 2009, Syron-Jones and Macintyre 2021). A systematic review of four randomized controlled trials reported a tendency to lower scar height in scars treated with MCGs (Anzarut et al. 2009). However, some researchers question the efficacy of MCGs in treating burn scars (Ai et al. 2017), and it is difficult to draw conclusions, due to the heterogeneity of burn injuries (depth, anatomical site) and pressure levels applied (Macintyre and Baird 2006).

There is confusion about what constitutes safe and effective pressure levels, and a major limitation of existing research is that compression is rarely measured by researchers (Macintyre 2007), and when it has been measured, it has been shown to be varied (e.g., Giele et al. 1997, Macintyre and Baird 2006, Wiseman et al. 2019). Some researchers have reported discomfort (Chan and Fan 2002) and potentially harmful effects of high-pressure levels that can cause deformity (Harumi et al. 2001, Miyatsuji et al. 2002, Macintyre and Baird 2006). There is no scientific evidence for an optimum pressure level for burn scar treatments (Macintyre and Baird 2006, Wiseman et al. 2019); nevertheless some researchers suggest pressure levels of 24/25 mmHg, as this equals capillary pressure (Williams, Knapp and Wallen 1998, Macintyre and Baird 2006). There have also been suggestions of 15 mmHg as an effective compression level that is tolerated by patients (Ai et al. 2017), leading to a consensus of 15–25 mmHg as a suitable pressure range for burn scar treatments (Wiseman et al. 2019).

11.2.3 Comments and Recommendations for Further Research

Even though compression therapy has been established for decades, the efficacy of treating medical conditions has not been unequivocally proven. Many existing studies have a small sample size and poor methodology. There are many unanswered questions when it comes to the optimal compression treatment for specific conditions. Without fully understanding the underlying mechanisms of compression therapy for venous disorders and burn scars, it is difficult to prove their efficacy scientifically and provide recommendations for application regimes and optimal pressure levels. More large-scale, randomized control trials are needed to evidence the effects of CGs. Future studies should clearly outline a rigorously designed experimental protocol including devices, techniques, and potential biases (Mosti and Cavezzi 2019). To better understand the effects of compression, it is critical to measure the interface pressure, as it is the dosage of compression therapy.

11.3 EFFECTIVENESS OF COMPRESSION GARMENTS IN SPORTS

Recreational and elite athletes wear SCGs for training, recovery, and in many disciplines also competition purposes, as SCG companies promise to make them "better athlete[s] in both performance and recovery" (Skins 2022) with "a reduced risk of injury" (2XU 2022). With the increasing popularity of SCGs, the number of studies assessing the efficacy of SCGs has substantially increased over the past two decades. However, despite many existing studies, the performance- and recovery-enhancing properties promised by SCG brands have not been unequivocally proven.

Most studies assessing the effects of SCGs on exercise performance or recovery measure various parameters before, during, and/or after exercise for a control condition and the SCG condition. Researchers have used many different mechanical and physiological parameters to assess SCG efficacy, a list of the most common parameters is given in Table 11.1.

TABLE 11.1
Examples of Mechanical and Physiological Parameters Measured to Assess Sports Performance and Recovery

	Mechanical parameters	Physiological parameters
Performance	• Exercise time • Distance covered • Total workout • Maximum voluntary force generation • Time to exhaustion	• Blood pH • Oxygenation • VO_2 (max) • VO_2 (peak) • Heart rate • Hemoglobin
Recovery	• Subsequent workout performance parameters (see above) • Muscle swelling	• Lactate • Creatine kinase • C-reactive protein • Body temperature

11.3.1 Evidence for the Effects of Compression Garments on Exercise Performance

The exact mechanisms of SCGs are not fully understood, although it is likely that they differ during various exercise types (e.g., strength, endurance), owing to the different physical conditions and metabolic systems used. Researchers (Born, Sperlich and Holmberg 2013) have suggested five different categories of effects of SCGs on the human body: hemodynamic, mechanical, neuromuscular, thermal, and psychological effects.

11.3.1.1 Hemodynamic/Cardiodynamic Effects

During exercise, especially endurance exercise, there are increasing demands on the cardiodynamic system to provide adequate cardiac output and venous return (Couturier and Duffield 2013). Zaleski et al. (2015) reported that compression socks can reduce hemostatic activation in marathon runners; however, other studies have been inconclusive (Born et al. 2014). SCGs had little effect on participants' heart rates during various exercise protocols (e.g., Bringard, Perrey and Belluye 2006, Duffield and Portus 2007, Leicht, Ahmadian and Nakamura 2020, Gimenes et al. 2021), although conversely a small number of studies reported positive effects of compression on heart rate (e.g., Dascombe et al. 2011, Broatch et al. 2019, Toolis and McGawley 2020). A slower heart rate is the result of increased heart stroke volume caused by enhanced venous return (Born et al. 2014); thus, it is unclear if SCGs have significant effects on the venous return of healthy athletes during exercise.

11.3.1.2 Mechanical Effects

The mechanical effects of wearing SCGs are mainly based on the notion of providing support to the soft tissue and reducing muscle vibration. When muscles oscillate, fiber recruitment increases as muscles attempt to dampen the oscillation (Wakeling, Jackman and Namburete 2013), which increases energy expenditure. Reducing muscle vibration thus could result in longer time to fatigue (Born, Sperlich and Holmberg 2013), but so far there is only limited evidence supporting this effect (Ehrstrom et al. 2018).

11.3.1.3 Neuromuscular Effects

There is limited existing research measuring the neuromuscular effects of SCGs. Kraemer et al. (1998) reported enhanced proprioception resulting in improved joint position sense of the hip, which increased power endurance during repetitive jumps. While improvements in joint proprioception were also found by other researchers (Xhang et al. 2019, Cheng and Xiong 2019), the throwing accuracy of cricketers was not improved by wearing three different whole-body SCGs (Duffield and Portus 2007). Lee and colleagues (2017) reported that compression affects the nervous system, increasing agility through enhanced motor-related information processing and focus, based on their findings measuring brain waves during side-step tests.

Proprioceptive effects are difficult to measure due to the complexity of muscle receptors engaging in proprioception (Bernhardt and Anderson 2005), the potential subtlety of effects (Hooper et al. 2015), and the likely interrelations of different compression effects. More evidence is needed to draw conclusions on the neuromuscular effects of SCGs.

11.3.1.4 Thermal Effects

There is mixed evidence on the thermal effects of SCGs. Skin temperature increased in several (e.g., Duffield and Portus 2007, MacRae et al. 2012, Priego Quesada et al. 2015), but not all (Leoz-Abaurrea, Tam and Aguado-Jimenez 2016), studies using a variety of exercise protocols; however, core temperature was not affected, and it appears that increased skin temperature had no negative effect on performance in existing SCG endurance studies reviewed by Engel et al. (2016).

11.3.1.5 Psychological Effects

SCGs could have psychological effects on wearers. A sensation of comfort during exercise and a positive attitude toward SCGs can contribute to enhanced performance (Hooper et al. 2015, Brophy-Williams et al. 2017). However, these effects are difficult to measure, and the extent to which positive effects of SCGs found in the literature are based on a placebo effect is unknown.

Several researchers, using various exercise protocols, assessed the effect of SCGs on ratings of perceived exertion. Most studies (e.g., Ali, Caine and Snow 2007, Leicht, Ahmadian and Nakamura 2020) found no effect on perceived exertion, whereas only a few researchers reported reduced perceived exertion when wearing SCGs (Goh et al. 2011, Rugg and Sternlicht 2013). It has been recommended to consider the participants' beliefs in the efficacy of SCGs (Brophy-Williams et al. 2019), which has been neglected by most researchers.

Overall, there is little evidence that SCGs enhance exercise performance. This is supported by several review studies that reported no or minimal effects of SCGs on exercise performance (MacRae et al. 2012, Born, Sperlich and Holmberg 2013, Engel et al. 2016, da Silva et al. 2018, Weakley et al. 2021). However, at the same time, no evidence suggests adverse effects of wearing SCGs during exercise and a potential placebo effect could boost wearers' performance.

11.3.2 Evidence for the Effects of Compression Garments on Post-exercise Recovery

Managing post-exercise recovery is an important element of training and can affect subsequent exercise performance. Athletes wear SCGs during and/or after exercise in the hope of reducing the severity and time frame of symptoms, such as delayed onset muscle soreness (DOMS), which is particularly important during competitions (Bringard, Perrey and Belluye 2006). Researchers have suggested that there is more evidence for the beneficial effects of SCGs on recovery than for performance (e.g., Beliard et al. 2015), and several review studies (Hill, Howatson, van Someren, Leeder, et al. 2014, Marques-Jimenez et al. 2016, Brown et al. 2017, da Silva et al. 2018, Lee et al. 2022) concluded that wearing SCGs has some positive effects on aspects of recovery. However, study designs of research measuring the effects of SCGs on recovery vary even more than studies focusing on performance effects. There are particularly large differences between wear times, which range from one hour (e.g., Driller and Halson 2013) to several days (e.g., Kraemer et al. 2001) post-exercise. It is currently not clear when (i.e., during exercise, after exercise, or both) and for what length of time (e.g., 12 h, 24 h, 48 h) SCGs should be worn for optimal recovery. Consequently, findings of existing research vary, while some researchers found improvements in physiological and perceptual markers of post-exercise damage (e.g., Duffield and Portus 2007, Driller and Halson 2013), others only found improved subjective perceptions of recovery (Pruscino, Halson and Hargreaves 2013, Hill, Howatson, van Someren, Walshe, et al. 2014). Similarly, SCG use resulted in improved subsequent endurance performance in some studies (e.g., De Glanville and Hamlin 2012), but others reported no effects (Goto, Mizuno and Mori 2017).

Engel et al. (2016) concluded that SCGs can reduce recovery symptoms when worn during and after intense or prolonged endurance exercise, likely due to improved blood flow and associated removal of metabolic waste. It has also been suggested that compression reduces the physical space for soft tissue to swell or hemorrhage (French et al. 2008) and that reducing muscle vibration could minimize structural muscle damage, resulting in reduced DOMS (Ali, Caine and Snow 2007). However, the underlying mechanism and responses of wearing SCGs are not fully understood. Even though there is a tendency for positive effects of SCGs on recovery reported by several researchers, there is still no unambiguous evidence to confirm that wearing SCGs improves recovery outcomes. The small sample sizes and heterogeneity of study designs particularly in terms of wear times, pressure levels, and training status mean studies cannot be directly compared.

11.3.3 Pressure Levels of Sports Compression Garments

The purpose of SCGs is to apply pressure to the underlying body. However, when reviewing existing studies, there is no recognizable trend in pressure levels that lead to positive results for performance or recovery over other pressure levels. Vast variations in pressure levels exist across studies and pressure ranges resulting in positive effects and no effects overlap. The use of different pressure levels in the same exercise protocol also resulted in mixed findings (e.g., MacRae et al. 2012, Miyamoto and Kawakami 2014, Zinner et al. 2017), providing no insight into minimum or optimal pressure levels.

One of the main concerns with existing studies on the efficacy of SCGs is that the proportion of studies measuring compression in vivo across multiple locations for each study participant is small. There could be a wide range of pressures exerted among participants of studies that report estimated pressures due to different limb girths and tissue compositions (MacRae, Cotter and Laing 2011, Hill et al. 2015). It is, therefore, essential that future studies report pressure levels measured in vivo using a reliable pressure measurement device to enable interpretation of the research and to develop a better understanding of the required pressure levels and effects of wearing SCGs (MacRae, Laing and Partsch 2016). Pressure should be understood as the dosage of compression treatment, which is a critical variable that needs to be reported in the same way as the dosage of, for instance, nutritional supplements would be reported.

Further research is needed to identify optimal pressure levels for different compression-related effects and wear environments (e.g., exercise type, location on body) (MacRae, Laing and Partsch 2016). Without knowing what the optimal pressure for specific outcomes and applications is, it is difficult to use SCGs as an effective ergogenic or recovery aid.

11.3.4 Comments and Recommendations for Further Research

Different exercise types use different energy systems and muscles, thus they have fundamentally different effects on the human body. Also, the location of pressure application (full body, full leg, thighs only, calves only, etc.) and time of application are likely to influence the effects of SCGs. Due to the heterogeneity of studies (Table 11.2) and the "fuzziness" of many study designs (e.g., garment type, pressure level, fit on body), no meaningful conclusions can be drawn.

One of the problems with trying to quantify the effects of SCGs on exercise performance and recovery is the difficulty to eliminate a potential placebo effect. A sensation of vitality can significantly affect exercise performance and feelings of exertion. SCGs can therefore act as an ergogenic aid even in the absence of any physiological effects (Born, Sperlich and Holmberg 2013). As it is unlikely that participants can be fully blinded due to the obvious difference in the tightness of compression and placebo garments (MacRae, Cotter and Laing 2011), participants' beliefs in the functionality of SCGs and previous experience with SCGs should be recorded (Brophy-Williams et al. 2019, Gimenes et al. 2021). There is limited research that explores users' perceptions and attitudes toward SCGs.

TABLE 11.2
Heterogeneity of Existing Studies on the Functionality of SCGs

	Heterogeneity
Participants	• Sex • Age • Level of training • Familiarity with exercise protocol • BMI • Muscle mass • Body surface tissue composition
Exercise protocol	• Type of exercise (e.g., endurance vs. resistance) • Intensity and duration of exercise • Time of test • Environmental conditions
SCG intervention	• Garment type/body coverage • Wear time • Wear duration • Garment design (e.g., fabric, seams, special features) • Levels of pressure applied by SCGs
Measuring effects of SCGs	• Performance and recovery variables measured • Interpretation of data

With research mainly conducted in the fields of physiology, biochemistry, biomechanics, and sports medicine (Fu, Liu and Fang 2013), considerations of garment aspects (e.g., fabric properties, garment design, and construction) have been neglected. It is critical that researchers focus more on the SCG and describe at least the basic characteristics of the garments (type, body coverage, brand, material composition) and the way they were fitted. Any obvious fit issues of the garments should also be reported, as these could affect the reproducibility of studies.

It is currently unclear if SCGs have any beneficial effects on exercise performance or recovery; however, it appears that wearing SCGs during or after exercise has no significant detrimental effects on the wearer's health and exercise outcomes (Weakley et al. 2021). There is a need for studies with larger sample sizes and a variety of different training statuses for various exercise types. These studies should also measure perceptual effects and participants' beliefs in SCGs in addition to objective performance markers. Even though a wealth of research on SCGs has emerged over the past 20 years, the interdisciplinarity of the subject and complexity of the relationship between SCGs and the human body means that there are still many unanswered questions. A more standardized approach to the performance testing of SCGs with clear study designs is needed including a more detailed focus on the intervention: the SCG used and its properties and applied pressures.

11.4 CONCLUSIONS

This chapter provided a brief overview of evidence found in existing research that evaluates the efficacy of CGs in medical and sports applications. CGs have been established in the medical field for many decades and are the cornerstone in the treatment of DVT, venous ulcers, edema, and burn scars. CGs have been shown to improve venous disorders, but the optimal application regime and pressure magnitude are not scientifically founded. While there is clinical evidence of the efficacy of CGs in treating burn scars, scientific evidence is currently ambiguous. There are still many areas of compression therapy that are unclear, including the underlying mechanisms for the treatment of different conditions, which makes it difficult to evidence efficacy.

With the increasing popularity of SCGs, the number of studies evaluating the functionality of SCGs has grown substantially. However, due to the heterogeneity and small sample sizes of the studies, it is impossible to draw conclusions on whether wearing SCGs can enhance sports performance and post-exercise recovery. In addition, details such as optimal pressure level and application site are currently elusive and need to be established for specific applications. Larger research studies with clearly described methodologies are needed to evidence the effects and underlying mechanisms causing the effects of wearing CGs for different applications in medical and sports settings. Overall, researchers have neglected garment fit, and many have not measured in vivo compression. No attention has been paid to considerations of the whole complexity of CGs, such as the relationships between applied pressure, body dimensions, movement, comfort, and garment characteristics. Interdisciplinary teams understanding the human body, exercise science, textile materials, and garment construction are best suited to conduct this research.

REFERENCES

2XU. 2022. "Compression". https://uk.2xu.com/collections/compression.

Agu, Obi, Hamilton, G, and Baker, D. 1999. "Graduated compression stockings in the prevention of venous thromboembolism". *British Journal of Surgery* 86 (8):992–1004.

Ai, Jin-Wei, Liu, Jiang-tao, Pei, Sheng-Duo, Liu, Yu, Li, De-Shaeng, and Lin, Hong-ming. 2017. "The effectiveness of pressure therapy (15-25 mmHg) for hypertrophic burn scars: A systematic review and meta-analysis". *Scientific Reports* 7 (40185):1–11.

Ali, A, Caine, M P, and Snow, B G. 2007. "Graduated compression stockings: Physiological and perceptual responses during and after exercise". *Journal of Sports Science* 25 (4):413–419. doi: 10.1080/02640410600718376.

Amsler, Felix, Torsten, Willenberg, and Werner, Blättler. 2009. "In search of optimal compression therapy for venous leg ulcers: A meta-analysis of studies comparing diverse [Corrected] Bandages with specifically designed stockings". *Journal of Vascular Surgery* 50 (3):668–674.

Anzarut, A, Olson, J, Singh, P, Rowe, B H, and Tredget, E E. 2009. "The effectiveness of pressure garment therapy for the prevention of abnormal scarring after burn injury: A meta-analysis". *Journal of Plastic, Reconstructive & Aesthetic Surgery* 62 (1):77–84.

Ashby, Rebecca L, Gabe, R, Ali, S, Adderley, U, et al. 2014. "Clinical and costeffectiveness of compression hosiery versus compression bandages in treatment of venous leg ulcers (venous leg ulcer study iv, venus iv): A randomised controlled trial". *Lancet* 383 (9920):871–879.

Beliard, Samuel, Michel, Chauveau, Timothee, Moscatiello, Francois, Cros, Fiona, Ecarnot, and Francois, Becker. 2015. "Compression garments and exercise: No influence of pressure applied". *Journal of Sports Science and Medicine* 14 (1):75–83.

Berman, B, and Flores, F. 1998. "The treatment of hypertrophic scars and keloids". *European Journal of Dermatology* 8 (8):591–596.

Bernhardt, Theresa, and Gregory, S Anderson. 2005. "Influence of moderate prophylactic compression on sport performance". *Journal of Strength and Conditioning Research* 19 (2):292–297.

Best, A J, Williams, S, Crozier, A, Bhatt, R, Gregg, P J, and Hui, A C W. 2000. "Graded compression stockings in elective orthopaedic surgery". *The Journal of Bone and Joint Surgery* 82 (B):116–118.

Blättler, W, and Partsch, H. 2003. "Leg compression and ambulation is better than bed rest for the treatment of acute deep venous thrombosis". *International Angiology* 22 (4):393–400.

Bochmann, Rolf P, Seibel, Woldemar, Haase, Elke, Hietschold, Volker, Rödel, Hartmut, and Deussen, Andreas. 2005. "External compression increases forearm perfusion". *Journal of Applied Physiology* 99 (6):2337–2344.

Born, Dennis-Peter, Hans-Christer, Holmberg, Florian, Goernert, and Billy, Sperlich. 2014. "A novel compression garment with adhesive silicone stripes improves repeated sprint performance – A multi-experimental approach on the underlying mechanisms". *BMC Sports Science, Medicine, and Rehabilitation* 6 (21):1–9.

Born, Dennis-Peter, Billy, Sperlich, and Hans-Christer, Holmberg. 2013. "Bringing light into the dark: Effects of compression clothing on performance and recovery". *International Journal of Sports Physiology and Performance* 8 (1):4–18.

Bringard, A, Perrey, S, and Belluye, N. 2006. "Aerobic energy cost and sensation responses during submaximal running exercise – Positive effects of wearing compression tights". *International Journal of Sports Medicine* 27 (5):373–378.

Broatch, J R, David, J Bishop, Emma, K Zadow, and Shona, Halson. 2019. "Effects of sports compression socks on performance, physiological, and hematological alterations after long-haul air travel in elite female volleyballers". *Journal of Strength and Conditioning Research* 33 (2):492–501.

Brophy-Williams, Ned, Matthew, W Driller, Kitic, C M, Fell, J W, and Halson, S L. 2017. "Effect of compression socks worn between repeated maximal running bouts". *International Journal of Sports Physiology and Performance* 12 (5):621–627.

Brophy-Williams, Ned, Matthew, W Driller, Kitic, C M, Fell, J W, and Halson, S L. 2019. "Wearing compression socks during exercise aids subsequent performance". *Journal of Science and Medicine in Sport* 22 (1):123–127.

Brown, Freddy, Conor, Gissane, Glyn, Howatson, Ken van, Someren, Charles, Pedlar, and Jessica, Hill. 2017. "Compression garments and recovery from exercise: A meta-analysis". *Sports Medicine* 47 (11):2245–2267. doi: 10.1007/s40279-017-0728-9.

Buhs, Chad L, Phillip, J Bendick, and John, L Glover. 1999. "The effect of graded compression elastic stockings on the lower leg venous system during daily activity". *Journal of Vascular Surgery* 30 (5):830–835.

Chan, A P, and Fan, J. 2002. "Effect of clothing pressure on the tightness sensation of girdles". *International Journal of Clothing Science and Technology* 14 (2):100–110. doi: 10.1108/09556220210424215.

Cheng, Longfei, and Caihua, Xiong. 2019. "The effects of compression stockings on the energetics and biomechanics during walking". *European Journal of Applied Physiology* 119:2701–2710.

Choucair, Michelle, and Tania, J Phillips. 1998. "Compression therapy". *Phlebology* 24:141–148.

Couturier, Antoine, and Rob, Duffield. 2013. "Compression garments". In *Recovery for Performance in Sport*, edited by Christophe Hausswirth and Iigo Mujika, 135–142. Champaign, IL: Human Kinetics.

Dascombe, Ben J, Trent, K Hoare, Joshua, A Sear, Peter, R Reaburn, and Aaron, T Scanlan. 2011. "The effects of wearing undersized lower-body compression garments on endurance running performance". *International Journal of Sports Physiology and Performance* 6 (2):160–173.

De Glanville, Kieren M, and Michael, J Hamlin. 2012. "Positive effect of lower body compression garments on subsequent 40-km cycling time trial performance". *Journal of Strength and Conditioning Research* 26 (2):480–486.

Dennis, M, Sandercock, P A, Reid, J, Graham, C, Murray, G, Venables, G, Rudd, A, and Bowler, G. 2009. "CLOTS trials collaboration, effectiveness of thigh-length graduated compression stockings to reduce the risk of deep vein thrombosis after stroke (CLOTS Trial 1): A multicentre, randomised controlled trial". *Lancet* 373:1958–1965.

Dissemond, Joachim, Bernd, Assenheimer, Bultemann, Anke, Gerber, Veronika, Gretener, Silvia, Kohler-von Siebenthal, Elisabeth, Koller, Sonja, et al. 2016. "Compression therapy in patients with venous leg ulcers". *Journal of the German Society of Dermatology* 14 (11):1072–1087.

Driller, Matthew W, and Shona L Halson. 2013. "The effects of lower-body compression garments on recovery between exercise bouts in highly-trained cyclists". *Journal of Science and Cycling* 1 (2):45–50.

Duffield, R, and Portus, M. 2007. "Comparison of three types of full-body compression garments on throwing and repeat-sprint performance in cricket players". *British Journal of Sports Medicine* 41 (7):409–414. doi: 10.1136/bjsm.2006.033753.

Ehrstrom, Sabine, Mathieu, Gruet, Marlene, Giandolini, Serge, Chapuis, Jean-Benoit, Morin, and Fabrice, Vercruyssen. 2018. "Acute and delayed neuromuscular alterations induced by downhill running in trained trail runners: Beneficial effects of high-pressure compression garments". *Frontiers in Physiology* 9 (1627):1–18.

Elwell, Rebecca. 2016. "An overview of the use of compression in lower-limb chronic oedema". *Nurse Prescribing* 14 (11):554–558.

Engel, Florian, Christian, Stockinger, Alexander, Woll, and Billy, Sperlich. 2016. "Effects of compression garments on performance and recovery in endurance athletes". In *Compression Garments in Sports: Athletic Performance and Recovery*, edited by Florian Engel and Billy Sperlich, 33–61. Cham: Springer.

Feist, William R, Dominic, Andrade, and Leonard, Nass. 2011. "Problems with measuring compression device performance in preventing deep vein thrombosis". *Thrombosis Research* 128 (3):207–209.

French, Duncan N, Kevin, G Thompson, Stephen, W Garland, Barnesm, Christopher A, Matthew, D Portas, Peter, E Hood, and Graeme, Wilkes. 2008. "The effects of contrast bathing and compression therapy on muscular performance". *Medicine & Science in Sports & Exercise* 40 (7):1297–1306.

Fu, Weijie, Yu, Liu, and Ying, Fang. 2013. "Research advancements in humanoid compression garments in sports". *International Journal of Advanced Robotic Systems* 10 (66):1–6. doi: 10.5772/54560.

Galanaud, J P, Laroche, J P, and Righini, M. 2013. "The history and historical treatments of deep vein thrombosis". *Journal of Thrombosis and Haemostasis* 11 (3):402–411.

Giele, H P, Liddiard, K, Currie, K, and Wood, F M. 1997. "Direct measurement of cutaneous pressures generated by pressure garments". *Burns* 23 (2):137–141. doi: 10.1016/S0305-4179(96)00088-5.

Gimenes, Samuel Valencia, Maocir, Marocolo, Larissa, Neves Pavin, Leandro, Mateus Pagoto Spigolon, Octavio, Barbosa Neto, Bruno, Victor Correa da Silva, Rob, Duffield, and Gustavo, Ribeiro da Mota. 2021. "Compression stockings used during two soccer matches improve perceived muscle soreness and high-intensity performance". *Journal of Strength and Conditioning Research* 35 (7):2010–2017.

Goh, S S, Laursen, P B, Dascombe, B, and Nosaka, K. 2011. "Effect of lower body compression garments on submaximal and maximal running performance in cold (10 degrees c) and hot (32 degrees c) environments". *European Journal of Applied Physiology* 111 (5):819–826. doi: 10.1007/s00421-010-1705-2.

Goto, Kazushige, Sahiro, Mizuno, and Ayaka, Mori. 2017. "Efficacy of wearing compression garments during post-exercise period after two repeated bouts of strenuous exercise: A randomized crossover design in healthy, active males". *Sports Medicine – Open* 3 (1):1–10. doi: 10.1186/s40798-017-0092-1.

Harumi, M, Miyuki, N, Hideo, M, and Kiyokazu, K. 2001. "Effects of clothing pressure exerted on a trunk on heart rate, blood pressure, skin blood flow and respiratory function". *Journal of Textile Machinery Society of Japan* 54 (2):57–62.

Hill, Jessica A, Howatson, Glyn, van Someren, Ken, Davidson, Stuart, and Pedlar, Charles R. 2015. "The variation in pressures exerted by commercially available compression garments". *Sports Engineering* 18 (2):115–121. doi: 10.1007/s12283-015-0170-x.

Hill, Jessica A, Howatson, Glyn, van Someren, Ken, Leeder, J, and Pedlar, Charles R. 2014. "Compression garments and recovery from exercise-induced muscle damage: A meta-analysis". *British Journal of Sports Medicine* 48 (18):1340–1346. doi: 10.1136/bjsports-2013-092456.

Hill, Jessica A, Howatson, Glyn, van Someren, Ken, Ian, Walshe, and Pedlar, Charles R. 2014. "Influence of compression garments on recovery after marathon running". *Journal of Strength and Conditioning Research* 28 (8):2228–2235.

Hooper, David R, Lexie, L Dulkis, Paul, J Secola, Gabriel, Holtzum, Sean, P Harper, Ryan, J Kalkowski, Brett, A Comstock, et al. 2015. "Roles of an upper-body compression garment on athletic performance". *Journal of Strength and Conditioning Research* 29 (9):2655–2660.

Ibegbuna, Veronica, Konstantinos, T Delis, Andrew, N Nicolaides, and Olayide, Aina. 2003. "Effect of elastic compression stockings on venous hemodynamics during walking". *Journal of Vascular Surgery* 37 (2):420–425.

Junger, M P H, Ramelet, A, and Zuccarelli, F. 2004. "Efficacy of ready-made tubular compression device versus short-stretch compression bandages in the treatment of venous leg ulcers". *Wounds* 16 (10):313–320.

Kankariya, Nimesh, Laing, R M and Wilson, C A. 2021. "Textile-based compression therapy in managing chronic oedema: Complex interactions". *Phlebology* 36 (2):100–113.

Kim, Jayne Y, James, J Willard, Dorothy, M Supp, Sashwati, Roy, Gayle, M Gordillo, Chandan, K Sen, and Heather, M Powell. 2015. "Burn scar biomechanics after pressure garment therapy". *Plastic and Reconstructive Surgery* 136 (3):572–581.

Kraemer, William J, Bush, Jill A, Newton, Robert U, Duncan, Noel D, Volek, Jeff S, Denegar, Craig R, Canavan, Paul, Johnston, John, Putukian, Margot, and Sebastianelli, Wayne J. 1998. "Influence of a compression garment on repetitive power output production before and after different types of muscle fatigue". *Sports Medicine, Training and Rehabilitation* 8 (2):163–184.

Kraemer, William J, Bush, Jill A, Robbin, B Wickham, Craig, R Denegar, Ana, L Gomez, Lincoln, A Gotshalk, Noel, D Duncan, Jeff, S Volek, Margot, Putukian, and Wayne,

J Sebastianelli. 2001. "Influence of compression therapy on symptoms following soft tissue injury from maximal eccentric exercise". *Journal of Orthopaedic & Sports Physical Therapy* 31 (6):282–290.

Labarere, Jose, Bosson, Jean-Luc, Sevestre, Marie-Antoinette, Delmas, Anna-Sophie, Dupas, Stephane, Thenault, Marie-Helene, Legagneux, Annie, Boge, Gudrun, Terriat, Beatrice, and Pernod, Gilles. 2006. "Graduated compression stocking thromboprophylaxis for elderly inpatients a propensity analysis". *Journal of General Internal Medicine* 21 (12):1282–1287.

Lee, B B, Andrade, M, Antignani, P L, Boccardo, F, Bunke, N, Campisi, C, Damstra, R, et al. 2013. "Diagnosis and treatment of primary lymphedema. consensus document of the international union of phlebology (IUP)-2013". *International Angiology* 32 (6):541–574.

Lee, Daniel C W, Ajmol, Ali, Sinead, Sheridan, Derwin, KC Chan, and Stephen HS Wong. 2022. "Wearing compression garment enhances central hemodynamics? A systematic review and meta-analysis". *Journal of Strength and Conditioning Research* 36 (8):2349–2359.

Lee, H, Kim, K, and Lee, Y. 2017. "The effect of the pressure level of sports compression pants on dexterity and movement-related cortical potentials". *Science & Sports*. doi: 10.1016/j.scispo.2017.03.006.

Leicht, Anthony S, Mehdi, Ahmadian, and Fabio, Nakamura Y. 2020. "Influence of lower body compression garments on cardiovascular autonomic responses prior to, during and following submaximal cycling exercise". *European Journal of Applied Physiology* 120 (7):1601–1607. doi: 10.1007/s00421-020-04391-9.

Leoz-Abaurrea, Iker, Nicholas, Tam, and Roberto, Aguado-Jimenez. 2016. "Impaired cardiorespiratory responses when wearing an upper body compression garment during recovery in a hot environment (40 °C)". *The Journal of Sports Medicine and Physical Fitness* 56 (6):684–691.

Liu, Rong, Xia, Guo, Terence, T Lao, and Trevor, Little. 2017. "A critical review on compression textiles for compression therapy: Textile-based compression interventions for chronic venous insufficiency". *Textile Research Journal* 87 (9):1121–1141.

Macintyre, L. 2007. "Designing pressure garments capable of exerting specific pressures on limbs". *Burns* 33 (5):579–586. doi: 10.1016/j.burns.2006.10.004.

Macintyre, L, and Baird, M. 2006. "Pressure garments for use in the treatment of hypertrophic scars – A review of the problems associated with their use". *Burns* 32 (1):10–15. doi: 10.1016/j.burns.2004.06.018.

MacRae, Braid A, Cotter, J D, and Laing, R M. 2011. "Compression garments and exercise: Garment considerations, physiology and performance". *Sports Medicine* 41 (10):815–843. doi: 10.2165/11591420-000000000-00000.

MacRae, Braid A, Laing, R M, Niven, B E, and Cotter, J D. 2012. "Pressure and coverage effects of sporting compression garments on cardiovascular function, thermoregulatory function, and exercise performance". *European Journal of Applied Physiology* 112 (5):1783–1795. https://doi.org/10.1007/s00421-011-2146-2.

MacRae, Braid A, Laing, R M, and Partsch, H. 2016. "General considerations for compression garments in sports: Applied pressures and body coverage". In *Compression Garments in Sports: Athletic Performance and Recovery*, edited by Florian Engel and Billy Sperlich, 1–32. Switzerland: Springer International Publishing.

Marques-Jimenez, Diego, Calleja-Gonzalez, J, Arratibel, Inaki, Delextrat, Anne, and Terrados, Nicolas. 2016. "Are compression garments effective for the recovery of exercise-induced muscle damage? a systematic review with meta-analysis". *Physiology & Behavior* 153 (2016):133–148.

Miyamoto, N, and Kawakami, Y. 2014. "Effect of pressure intensity of compression short-tight on fatigue of thigh muscles". *Medicine & Science in Sports & Exercise* 46 (11):2168–2174. doi: 10.1249/MSS.0000000000000330.

Miyatsuji, A, Matsumoto, T, Mitarai, S, Kotabe, Tetsuro, Takeshima, Takehiro, and Watanuki, Shigeki. 2002. "Effects of clothing pressure caused by different types of brassieres on autonomic nervous system activity evaluated by heart rate variability power spectral analysis". *Journal of Physiological Anthropology and Applied Human Science* 21 (1):67–74.

Moffatt, Christine. 2008. "Variability of pressure provided by sustained compression". *International Wound Journal* 5 (2):259–265.

Mosti, Giovanni, and Attilio, Cavezzi. 2019. "Compression therapy in lymphedema: Between past and recent scientific data". *Phlebology* 34 (8):515–522. doi: 10.1177/0268355518824524.

O'Meara, A, Nicky, Cullum, Elizabeth, Andrea Nelson, and Dumville, J C. 2012. "Compression for venous leg ulcers". *The Cochrane Database of Systematic Reviews* 11:CD000265.

Partsch, Hugo. 2012. "Compression therapy: Clinical and experimental evidence". *Annals of Vascular Diseases* 5 (4):416–422.

Perrey, Stephane. 2008. "Compression garments: Evidence for their physiological effects". In *The Engineering of Sport 7*, edited by Margaret Estivalet and Pierre Brisson 2, 319–328. Paris: Springer-Verlag.

Priego Quesada, J I, Lucas-Cuevas, A G, Gil-Calvo, M, Gimenez, J V, Aparicio, I, Cibrian Ortiz de Anda, R Salvador, Palmer, Llana-Belloch, S, and Perez-Soriano, P. 2015. "Effects of graduated compression stockings on skin temperature after running". *Journal of Thermal Biology* 52:130–136.

Pruscino, C L, Halson, S, and Hargreaves, M. 2013. "Effects of compression garments on recovery following intermittent exercise". *European Journal of Applied Physiology* 113 (6):1585–1596. doi: 10.1007/s00421-012-2576-5.

Puzey, Ginny. 2001. "The use of pressure garments on hypertrophic scars". *Journal of Tissue Viability* 12 (1):11–15.

Rabe, Eberhard, Partsch, H, Hafner, Jürg, Lattimer, Christopher R, Mosti, Giovanni, Neumann, Martino, Urbanek, Tomasz, Huebner, Monika, Gaillard, Sylvain, and Carpentier, Patrick. 2018. "Indications for medical compression stockings in venous and lymphatic disorders: An evidence-based consensus statement". *Phlebology* 33 (3):163–184.

Ramelet, A A. 2002. "Compression therapy". *Dermatologic Surgery* 28 (1):6–10.

Rohan, Pierre-Yves, Pierre, Badel, Bertrand, Lun, Didier, Rastel, and Stephane, Avril. 2013. "Biomechanical response of varicose veins to elastic compression: A numerical study". *Journal of Biomechanics* 46 (3):599–603.

Rohan, Pierre-Yves, Pierre, Badel, Bertrand, Lun, Didier, Rastel, and Stephane, Avril. 2014. "Prediction of the biomechanical effects of compression therapy on deep veins using finite element modelling". *Annals of Biomedical Engineering* 43 (2):314–324.

Rugg, Stuart, and Eric, Sternlicht. 2013. "The effect of graduated compression tights, compared with running shorts, on counter movement jump performance before and after submaximal running". *Journal of Strength and Conditioning Research* 27 (4):1067–1073.

Sajid, Muhammad, Mital, Desai, Morris, Richard W, and George, Hamilton. 2012. "Knee length versus thigh length graduated compression stockings for prevention of deep vein thrombosis in postoperative surgical patients". *The Cochrane Database of Systematic Reviews* 5:CD007162.

Silva, César Augusto da, Lucas, Helal, Roberto, Pacheco da Silva, Karlyse, Claudino Belli, Daniel, Umpierre, and Ricardo, Stein. 2018. "Association of lower limb compression garments during high-intensity exercise with performance and physiological responses: A systematic review and meta-analysis". *Sports Medicine* 48 (8):1859–1873. doi: 10.1007/s40279-018-0927-z.

Skins. 2022. *Discover skins technology*. https://www.skinscompression.com/uk/discover-skins-technology/.

Syron-Jones, Dawn, and Lisa, Macintyre. 2021. "Pressure by design: How to improve the consistency of pressure garments in the clinical environment and implement a simple method for gathering evidence to establish efficacy". *Burns*, July. doi: 10.1016/j.burns.2021.07.019.

Toolis, Tom, and Kerry, McGawley. 2020. "The effect of compression garments on performance in elite winter biathletes". *International Journal of Sports Physiology and Performance* 16 (1):145–148.

Wade, Ros, Fiona, Paton, and Nerys, Woolacott. 2017. "Systematic review of patient preference and adherence to the correct use of graduated compression stockings to prevent deep vein thrombosis in surgical patients". *Journal of Advanced Nursing* 73 (2):336–348. doi: 10.1111/jan.13148.

Wakeling, James M, Meghan, Jackman, and Ana, I Namburete. 2013. "The effect of external compression on the mechanics of muscle contraction". *Journal of Applied Biomechanics* 29 (3):360–364.

Weakley, Jonathon, James, Broatch, Shane, O'Riordan, Matthew, Morrison, Nirav, Maniar, and Shona, L Halson. 2021. "Putting the squeeze on compression garments: Current evidence and recommendations for future research: A systematic scoping review". *Sports Medicine*, December. doi: 10.1007/s40279-021-01604-9.

Wells, P S, Lensing, A W A, and Hirsh, J. 1994. "Graduated compression stockings in the prevention of postoperative venous thromboembolism: A meta-analysis". *Archives of Internal Medicine* 154 (1):67–72.

Wienert, V. 2003. "Compression treatment after burns". In *Textiles and the Skin*, edited by P Elsner, K Hatch, and W Wigger-Alberti. Basel: Karger 31 (2003):108–113.

Williams, F, Knapp, D, and Wallen, M. 1998. "Comparison of the characteristics and features of pressure garments used in the management of burn scars". *Burns* 24 (4):329–335.

Wiseman, Jodie, Megan, Simons, Roy, Kimble, and Zephanie, Tyack. 2019. "Variability of pressure at the pressure garment-scar interface in children after burn: A pilot longitudinal cohort study". *Burns* 45 (1):103–113.

Xhang, Li Yin, Janos, Negyesi, Takeshi, Okuyama, Mami, Tanaka, Tibor, Hortobagyi, and Ryoichi, Nagatomi. 2019. "Position of compression garment around the knee affects healthy adults' knee joint position sense acuity". *Human Movement Science* 67 (2019):1–9.

Xiong, Ying, and Xiaoming, Tao. 2018. "Compression garments for medical therapy and sports". *Polymers* 10 (6). doi: 10.3390/polym10060663.

Yan, L H, Ping, L T W and Ping, Z Y. 2010. "Effect of different pressure magnitudes on hypertrophic scar in a chinese population". *Burns* 36 (8):1234–1241.

Zaleski, Amanda L, Kevin, D Ballard, Linda, S Pescatello, Gregory, A Panza, Brian, R Kupchak, Marcin, R Dada, William, Roman, Paul, D Thompson, and Beth, A Taylor. 2015. "The effect of compression socks worn during a marathon on hemostatic balance". *Physician and Sportsmedicine* 43 (4):336–341. doi: 10.1080/00913847.2015.1072456.

Zinner, Christoph, Maximilian, Pelka, Alexander, Ferrauti, Tim, Meyer, Mark, Pfeiffer, and Billy, Sperlich. 2017. "Responses of low and high compression during recovery after repeated sprint training in well-trained handball players". *European Journal of Sport Science* 17 (10):1304–1310. doi: 10.1080/17461391.2017.1380707.

12 Applications of Textile-Based Compression for Human Spaceflight and Extraterrestrial Surface Exploration

Braid A. MacRae, Jonathan Clarke, Abby Rudakov, James Waldie

CONTENTS

12.1 INTRODUCTION

Humans are more effective, fast, and capable during extraterrestrial surface exploration by several orders of magnitude over human-directed robots (Crawford 2012). Furthermore, humans have a key role in maintaining and repairing complex space hardware (Sullivan 2019). Execution of these tasks requires humans to be capable of effective work in the external space or extraterrestrial surface environments and to be able to successfully live and work long term inside spacecraft and surface habitats. Furthermore, those humans should be capable of a safe return to Earth.

DOI: 10.1201/9781003298526-12

Compression of the human body using textile materials has current and proposed uses for supporting such human endeavors in space. The objective of this chapter is to provide a general introduction to two applications of textile-based compression: (1) mechanical counterpressure (MCP) extra-vehicular activity (EVA) suits and (2) mechanical-loading countermeasure garments.

12.1.1 MCP EVA Suits

EVA involves an individual leaving the protection of a spaceship, habitat, or other pressurized vehicle, thereby entering the extreme-hypobaric or vacuum environment present in low Earth orbit or on the surface of the Moon or Mars (Table 12.1). Thus, to survive EVA, an EVA suit must meet several life-supporting requirements, one of which is the provision of pressure to the surface of the body. Traditionally, EVA suits have relied on gas pressurization, but, for reasons including lower mass and the potential for improved mobility, using textile-based compression (known in this context as MCP) has been proposed as an alternative. While gas-pressurized (GP) suits will likely be used for EVA upon initial return to the surface of the Moon (planned for the mid-2020s at the time of writing), MCP suits, or MCP-GP hybrids, provide compelling suit architectures for prolonged or subsequent surface exploration missions, particularly for Mars.

12.1.2 Mechanical-Loading Countermeasure Garments

The term countermeasure refers to an intervention for preventing or mitigating an unwanted action or outcome. The gravity on the surface of Mars and the Moon is much less than on Earth, and the body experiences microgravity onboard the international space station (ISS) and during transit on exploration missions (e.g., traveling to the Moon or Mars; Table 12.1). The prolonged unloading of the human body

TABLE 12.1
Absolute Pressure and Gravity at Selected Earth Altitudes, Low Earth Orbit (Orbit of the International Space Station), Moon Surface, and Mars Surface

Altitude or location	Absolute pressure		Gravity	
	kPa	atm	$m \cdot s^{-2}$	g
Earth sea level (0 km)	101.3	1.0	9.80	1.0
Beyond which supplementary body surface pressure is required (~13–15 km)	~16.4–11.7	~0.16–0.12	~9.76	~1.0
"Armstrong's line", where unconfined water boils at body temperature (~19 km)	~6.3	~0.06	~9.75	~0.99
International Space Station; low Earth orbit (~405 km) [a]	$2 \cdot 10^{-3}$	$2 \cdot 10^{-5}$	Actual: ~8.67 Effective: ~0	Actual: ~0.88 Effective: μg
Moon surface	$3 \cdot 10^{-10}$	$3 \cdot 10^{-15}$	1.62	0.17
Mars surface	0.6	$6 \cdot 10^{-3}$	3.71	0.38

[a] Inside the International Space Station is pressurized to be equivalent to Earth sea level. Microgravity (μg) is experienced due to continuous free fall during orbit, rather than a lack of Earth gravity per se.

in these reduced-gravity environments, in turn, drives physiological responses and adaptations that can be problematic for re-exposure to higher gravity. Specialized compression garments have existing and proposed roles as countermeasures for supporting physiological function during transient stress (e.g., orthostatic intolerance garments) and in preventing deconditioning that arises from a lack of body weight and associated axial load (countermeasure suits to 're-weight' the body).

12.2 EVA SUITS USING COMPRESSION TEXTILES

12.2.1 Compression for Very Low-Pressure or Vacuum Environments

The human body requires sufficient surface pressure to maintain physiological function, and, on Earth, this pressure is typically provided by the Earth's atmosphere. Atmospheric pressure results from the cumulative weight of gases comprising the atmosphere itself, and hence if a person was to ascend unprotected through Earth's atmosphere, they will be exposed to an ever-decreasing ambient pressure (Table 12.1). Ambient pressures beyond an altitude of ~13–15 km require assisted (positive pressure) breathing of 100% oxygen and supplementary pressure applied at the body surface. Beyond 18–19 km altitude (pressure <6.3 kPa), the body is effectively in space.[1]

The ambient environment outside the ISS and on the surface of the Moon is at or near vacuum (<0.002% of Earth's sea-level pressure), and while it does have an atmosphere, the ambient pressure on the surface of Mars is very low (<1% of Earth's). Accordingly, pressure suits covering the entire body are required to survive exposure to these foreign environments. Considering just the requirements of body surface pressure, an unprotected individual will otherwise experience the expansion of gas trapped in body cavities (e.g., intestinal tract and middle ear), rupture of capillaries in the skin, tissue swelling, boiling of exposed body water, formation of bubbles (principally nitrogen) in the body tissues and blood leading to decompression sickness and obstruction of perfusion, and ultimately loss of consciousness and death.

Most modern "full" pressure suits are pressurized when required with 100% oxygen, typically at a minimum of 25–30 kPa. In the context of spaceflight and surface exploration, there is a distinction between intra-vehicular activity (IVA) and EVA pressure suits. IVA suits are worn inside the spacecraft during phases like launch and re-entry and are intended to be pressurized only for short periods – long enough for the crew to be returned to a safe environment in the event of spacecraft cabin depressurization. As such, IVA suits generally need only allow for limited movement when pressurized and minimal independent oxygen supply (Jenkins 2012).

EVA suits, on the other hand, are pressurized for the full duration of use and, in association with an umbilical or backpack-mounted life support system, need to facilitate meaningful work in the external space or extraterrestrial surface environments.

12.2.2 MCP as an Alternative to GP EVA Suits

With increased government and private interest in crewed space exploration, humans will likely return to the surface of the Moon by the end of the 2020s and go on to

Mars in the following decades. Among the many challenges these explorers will face is having EVA suits that enable them to work safely and effectively on a *surface* with substantial *gravity* – conditions that impose different mobility and mass requirements versus more recent operational EVA suits used solely while 'floating' in microgravity.

To date, all EVA (Moon surface or microgravity) have been completed using GP EVA suits. 'Traditional' GP EVA suits are effective at providing the necessary pressure but, being gas pressurized, are extremely stiff and uncomfortable to move around in, requiring complex joints, significant exertion to carry out even simple tasks, and causing both minor and significant injuries (Chappell et al. 2017). Furthermore, GP EVA suits have been massive: almost 100 kg for an Apollo suit (Thomas and McMann 2012) and over 180 kg for the latest NASA prototype, the xEMU (Abercromby et al. 2020, NASA Office of Inspector General 2021).

A surface EVA suit must allow the wearer to effectively move and walk in a substantial gravity: 17% and 38% of Earth's for the Moon and Mars, respectively (Table 12.1). Therefore, overall suit mass is a major operational constraint for surface use, perhaps needing to be no more than ~170% of the body mass of the wearer on the Moon and ~50% on Mars (Wickman and Luna 1996). These percentages equate to ~119 kg and ~97 kg for the representative man and woman, respectively, on the Moon and a mere ~35 kg and ~28.5 kg, respectively, on Mars (Kent 2007). Surface EVA suits will need to support operations at distances of several kilometers from safety and must be comfortable and safe enough to enable EVAs every other day for weeks or even months (Drake 2009), equating to as many as several hundred uses on a 540-day Mars surface mission. Thus, a fundamental shift in EVA suit design and construction is required if a high EVA tempo is desired on the Moon and if productive manned Mars surface missions are indeed to follow.

An alternative EVA suit approach is to use direct mechanical loading of the body surface via the materials and fabrics of the suit itself. Such suits, known as MCP suits, arguably show more promise for being sufficiently flexible and lightweight to achieve intensive EVA goals, particularly for Mars. That said, considerable work remains to be done before MCP suits are a feasible replacement for GP suits (McFarland, Ross and Saunders 2019, McFarland 2022).

12.2.3 MCP EVA Suits

For a MCP suit, compression layers, often composed of elastic knitted fabrics, cover the entire body excluding the head and neck. The compression layers must provide acceptably uniform pressure over the body surface similar to that supplied by a GP suit (25–30 kPa) and are typically designed to be worn in combination with underlying comfort and slip layers, the latter to assist in donning. Common to prototype or conceptual MCP suits, compression gloves and socks are separate pieces, and a GP helmet is used and connected to a volume-compensating breathing bladder. This bladder sits under the compression layers on the torso to compensate for the change in volume of the chest with respiration, otherwise inspiration, and associated expansion of the chest, would increase the fabric tension and, thereby, local pressure (and vice versa with expiration). An environmental protection garment would be required over the top of the compression layer and conventional field boots may be worn for surface EVAs. In areas where full compression is difficult to achieve by fabric tension alone due to complex body curvature, the compression suit may be supplemented by gel or pneumatic packs or small bladders.

Plausible advantages of using compression textiles include that local failure of a MCP suit, such as puncture or tear, would not be almost immediately fatal; in the case of a GP suit, a tear would likely lead to widespread depressurization. MCP suits could allow the wearer to perform more difficult tasks and sustain a higher tempo of EVA operations as a result. Intrinsically simpler in design and construction than GP suits, a MCP suit might require less maintenance and would be more compact for storage.

Perhaps the most complete prototype MCP suits were constructed by a team led by Paul Webb in the 1960s and tested in a high-altitude chamber (Webb and Annis 1967, Annis and Webb 1971). Several different configurations of these suits were trialed, but all variations relied on a layered approach for reaching target pressures (Figure 12.1). This work demonstrated MCP suits could be much lighter and more flexible than GP suits, meaning that less effort was needed to work in them, with a

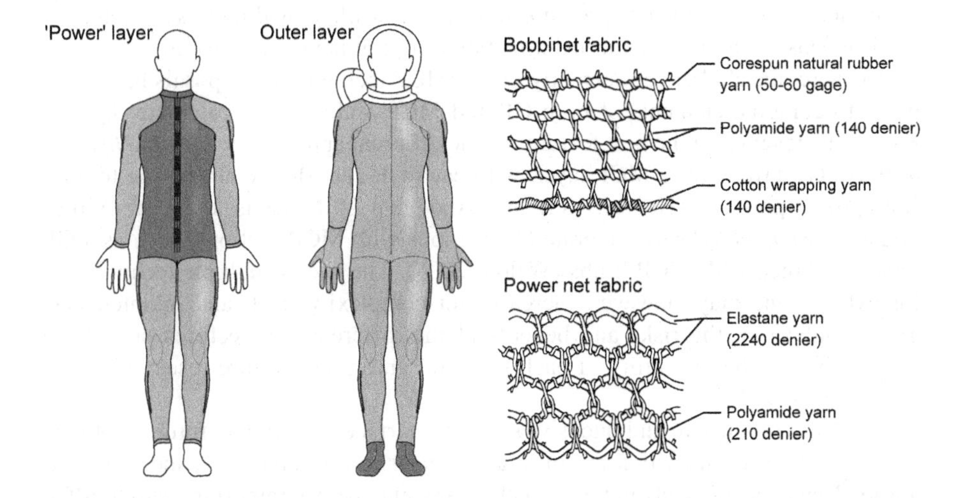

FIGURE 12.1 The "Space Activity Suit" – a prototype mechanical counterpressure suit. Two of the suit's constituent layers are depicted on the left, and a schematic of two key fabrics is shown on the right. Several so-called "power" compression layers were used comprising of power net fabric over the large circumferences of the torso (relative higher tension required for a given pressure) and bobbinet over the limbs (relative lower tension required for a given pressure); the outer compression layer was almost entirely comprised of bobbinet fabric (see Annis and Webb 1971, McFarland, Ross and Saunders 2019). Note that these suits would also require a protective layer (not shown) to be worn over the top of the outer compression layer, which would likely disrupt the suit's almost heroic silhouette. (Fabric schematics have been redrawn from Annis and Webb, 1971.)

flow-on reduction in the mass of oxygen and power for life support, cooling requirements, and potentially greater comfort.

Current MCP suits have low technology readiness levels of 3 (analytical and experimental critical function and/or characteristic proof-of-concept) (McFarland 2022), and no full-body MCP suit has been tested in a hypobaric chamber since the work of Annis and Webb (1971). Subsequent work has focused on components such as gloves (Waldie et al. 2002, Southern and Moiseev 2017), limbs (Newman, Canina and Trotti 2007), helmets (Kracik et al. 2012), suit construction (Jessup 2015), active tensioning (Holschuh and Newman 2016), thermal management (Stroming and Newman 2020b), and breathing bladders (Stroming and Newman 2020a). Technology gaps including sufficiently even and consistent pressure application, donning/doffing ease, and thermal regulation need to be addressed before MCP suits can be used operationally (McFarland 2022). There have been seemingly few recent works on the development of fabrics for EVA-specific requirements and applications, although whole-garment knitting of MCP components has been reported (Schauss et al. 2022).

A key distinction from conventional compression fabrics is the high stress required to deliver the high total pressures of 25–30 kPa (188–225 mmHg), particularly at sites with larger circumferences like the torso. Greater uncoupling of the circumferential and longitudinal fabric stresses (directions as worn on the body) under biaxial strain would further improve movement by enabling high stress in the circumferential direction (for pressure application) while providing less resistance to fabric extension in the longitudinal direction (for joint flexion/extension).

To overcome challenges currently facing MCP technology, proposals have been made to combine elements of both GP and MCP architectures to create 'hybrid' EVA suits, leveraging the best features of the different approaches. *Layered hybrids* utilize a GP layer over a MCP layer, with each GP and MCP contributing to total suit-applied pressure (Huerta, Kerr and Anderson 2018, Kluis and Diaz-Artiles 2021). *Segmented hybrids* integrate GP sections with MCP sections, such as a GP torso combined with MCP limbs (Waldie 2005). The use of multiple technologies for hybrid suits may, however, increase suit complexity, cost, and maintenance. Trade analysis of the risks and benefits of the different architectures provides a perspective on the feasibility of each proposal and guides future research efforts (Rudakov et al. 2022).

To ensure productive and intensive scientific surface exploration, future explorers will require EVA suits with lower mass and greater comfort, mobility, and maintainability. New developments in textile technology may be one important path to MCP EVA suit success.

12.3 COUNTERMEASURE GARMENTS USING COMPRESSION TEXTILES

12.3.1 The Need for Countermeasures

Gravity imposes a directional weight on the tissues and fluids comprising the human body. These static weights, and associated dynamic forces during movement, influence the structure and function of the body's organ systems. The reduction or absence of gravity with spaceflight and surface exploration unloads the body and results in deconditioning relative to the original Earth state.

Accordingly, a gamut of countermeasures is required for spaceflight and surface exploration missions. The intent when using countermeasures spans eliminating or preventing, but more commonly, mitigating adverse effects by retaining an appropriately conditioned state (e.g., inflight exercise, provision of gravity-like loading) and mitigating the adverse effects once deconditioning or maladaptation has occurred (e.g., fluid loading prior to reentry) (Clément 2011). Countermeasures therefore have a role at each stage of a mission, including before flight, inflight, and following flight.

The current countermeasure suite does a reasonable job for stays in microgravity onboard the ISS; however, there are challenges for surface exploration missions where volume and mass allowances are further constrained. For example, intensive aerobic and resistive exercise is a very important countermeasure onboard the ISS, but it is not clear if and what exercise equipment will be available for future exploration missions (Korth 2015). Textile-based compression has existing and proposed countermeasure roles and may have an increasing role in exploration missions to the Moon or Mars. Two applications considered here are orthostatic intolerance garments (OIG), worn when *leaving* microgravity, and countermeasure suits used to "re-weight" the body while *in* microgravity.

12.3.2 Orthostatic Intolerance Garments

Orthostatic intolerance (OI) is an umbrella term describing a disorder from being in an upright position, typically with respect to gravity, but vehicle accelerations or decelerations during launch or descent are also relevant. Hypotension (low blood pressure) and inadequate cerebral perfusion are important manifestations of OI, with symptoms including pre-syncope and syncope.

Interrelated factors associated with spaceflight likely contribute to postflight OI including reduced blood volume, ineffective blood vessel response during blood pressure challenges (blunted postural vasoconstrictor response), and deconditioning of the heart itself (Buckey et al. 1996, Clément 2011). The greatest risk of OI occurs upon return to Earth with the incidence after short duration (4–18 days) and long-duration spaceflight being in the order of 20% and 80%, respectively, and appears to disproportionately influence women (Nicogossian et al. 2016). The incidence also depends on the definition of OI used (Buckey et al. 1996) and likely also what countermeasures are used and what method is being used to determine OI (e.g., tilt or stand tests vs. during normal routine; duration of orthostatic stress). The risk of OI during a Moon or Mars descent/ascent is also relevant if constraints of the landing vehicle necessitate an upright body position (Lee et al. 2020). Further, even partial gravity like on Mars (0.38 *g*) might be enough to pose an OI risk during initial surface ambulation after prolonged microgravity exposure during transit (Lee et al. 2011).

Inflight countermeasures to postflight OI include exercise, lower-body negative pressure, and end-of-mission fluid loading. Suit cooling, recumbent posture, and specialized compression garments known as OIG are used during re-entry and landing. OIG also may be worn for hours or days during the return to terrestrial activities. In this way, OIG are worn to prevent OI once predisposing maladaptation to orthostatic stress has already occurred.

Effective OIG work by mechanically compressing the surface of the body which, in turn, compresses the tissues and vasculature, reducing venous pooling, promoting venous return, and maintaining arterial blood pressure. OIG typically cover the abdomen and lower limbs, with the abdominal region being the most important due to the high volume of blood in the splanchnic vascular bed (Rowell et al. 1972, Denq et al. 1997, Diedrich and Biaggioni 2004).

As an OI countermeasure during reentry and landing, shuttle-era NASA astronauts wore an air-pressurized (pneumatic) suit – the Anti-Gravity Suit (AGS) – covering the abdomen, thighs, and lower legs (Platts et al. 2009). While efficacious when inflated, limitations included that this suit restricted movement and ambulation and began to lose air pressure once disconnected from the air supply (Bishop et al. 1999, Lee et al. 2011).

The Russian OIG (Kentavr, or Centaur in English) is more akin to a terrestrial compression garment, relying on fabric tension in the garment itself for compression rather than air pressurization. NASA has transitioned to a similar approach for its contemporary OIG, a gradient compression garment (Stenger et al. 2013, Lee et al. 2020). Like the AGS, both these OIG target the vascular beds of the abdomen and lower limbs; NASA initially trialed a thigh-high only OIG but subsequently included the abdomen (Stenger et al. 2010).

The Kentavr is three separate pieces, has zippers to assist donning, and is adjustable using lacing down the sides (Figure 12.2); the nominal target pressure is 30 ± 5 mmHg (Vil-Viliams et al. 1998, Yarmanova et al. 2015). The Kentavr fabric has been reported as a knit ("stockinet") with bidirectional stretch, and as being "highly elastic" (Yarmanova et al. 2015). That said, the Russians did previously trial a version (Karkas) with only longitudinal stretch as worn on the body, and therefore stiff in the circumferential direction (Vil-Viliams et al. 1998). NASA's OIG is also three separate pieces for facilitating donning (Stenger et al. 2013; Figure 12.2). The knees and ankles are also covered (in part, to mitigate edema) and the target pressures are graduated from ~55 mmHg at the ankle to ~35 mmHg at the knee and to ~18 mmHg at the proximal thigh, and with ~16 mmHg over the abdomen (Stenger et al. 2014, Lee et al. 2020). Fiber content has been reported as nylon and elastane (Stenger et al. 2010). Little publicly reported information is available regarding the fabrics used in these Russian and American OIG. Similarly, data on the actual pressures measured while being worn are also rare, but are important (MacRae, Laing and Partsch 2016).

A challenge of relying on fabric tension for the compression from OIG (cf. pressurized air) is that local fabric extension in the donned state is critical for the applied pressure due to the underlying fabric stress–strain relationship. For these OIG, garment ease (i.e., fit) and, where applicable, to what extent they are tightened during adjustment are therefore critical for achieving the desired pressures. These considerations are important not only for the original measured body size and shape but particularly because body dimensions change with spaceflight (typically reduced

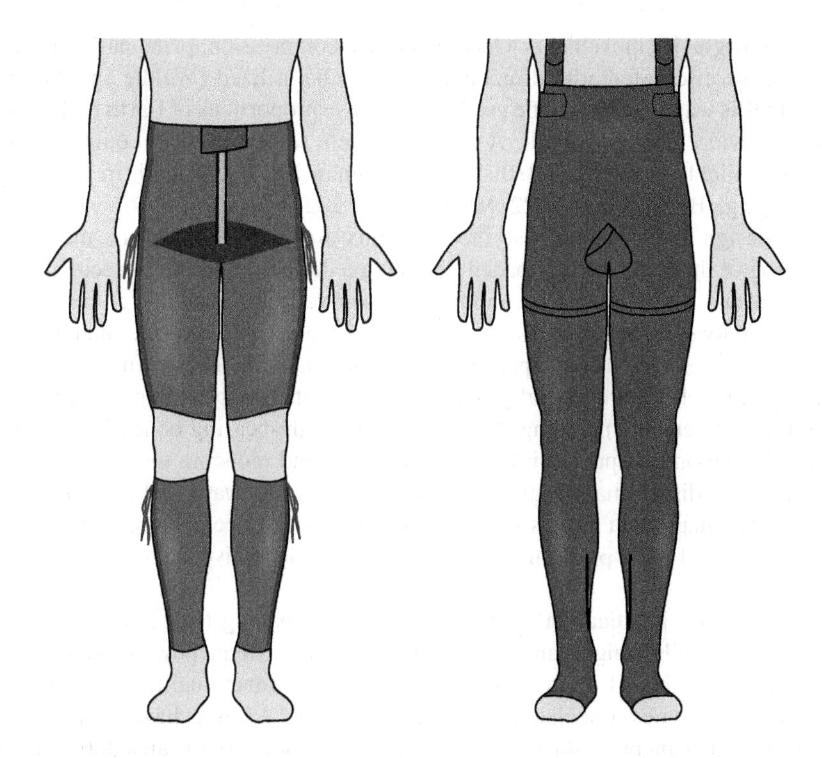

FIGURE 12.2 Russian (Kentavr; left) and US (gradient compression garment; right) orthostatic intolerance garments.

lower-body circumferences due to reduced lower-body fluid volume and muscular atrophy). Further challenges include that local pressures for a given fabric tension will depend on the local curvature for sites not sufficiently round in cross-section and that operational and time constraints mean steps like pressure measurement when donning are impractical. Utilizing "flatter" portions of the fabric stress–strain curve to minimize pressure changes with strain changes has been considered (Stenger et al. 2014) but requires validation.

Notwithstanding these challenges, advantages of fabric OIG include lower mass and volume, that the garment is easier to move in during pressure application, and that they can be worn for hours or days following landing (Bishop et al. 1999, Lee et al. 2011, Kozlovskaya, Yarmanova, and Fomina 2015, Lee et al. 2020). Future approaches for clarifying pressures, directly or indirectly, during adjustment will be valuable. Alternative approaches to improve comfort and donning have also been proposed (e.g., Granberry et al. 2020, Bunford et al. 2022).

12.3.3 Countermeasure Suits for Re-weighting the Body

Countermeasure suits for re-weighting the body are considered here as encompassing garments used to provide loading in the longitudinal body direction and to resist certain anti-gravity movements. General surface compression, principally via fabric strain in the circumferential direction, may also be utilized (Waldie and Newman 2011). In this way, textiles can be used to re-impose proportions of Earth body weight and to provide sensory stimuli. A more complete review of such countermeasure suits is available elsewhere (Bellisle and Newman 2020) and work in this area is ongoing (e.g., Bellisle et al. 2022, MacRae et al. 2022).

The design and proposed use of these suits are predicated on the mechanical unloading of the body during spaceflight being a driver of skeletal, muscular/neuromuscular, and sensorimotor deconditioning relative to the Earth (1 *g*) and perhaps partial-gravity (e.g., Moon or Mars) state. Accordingly, these suits are intended to be worn in microgravity during spaceflight but could also be used to further augment axial loading while in a partial gravity environment. Axial loading has a range of hypothesized benefits including preservation of weight-bearing bones, mitigation of spinal elongation (via preserving spinal curvature and reducing the swelling of the intervertebral discs), maintenance of neuromuscular pathways and muscular function, and, perhaps even more so in combination with surface compression, supporting awareness of body position and movement (proprioceptive and cutaneous tactile afferents).

Experimental justification for the magnitude of loading (e.g., full body weight versus partial body weight) and suit use protocols (e.g., frequency, duration) are not yet clear, although will likely depend on the intended functional targets. That said, a range of outcomes associated with suit-based axial loading have been reported including experimental (during hyper-buoyancy flotation) or anecdotal support for reduction or reversal of spinal elongation or stature increases (Kozlovskaya, Yarmanova and Fomina 2015, Yarmanova et al. 2015, Carvil 2017, Green and Scott 2018), changes in intervertebral disc biomechanics (modelling; Zhang et al. 2021), preservation of slow-twitch muscle contractile properties (Yamashita-Goto et al. 2001), and attenuation of spinal reflex hypersensitivity (Shigueva et al. 2015). Further, provision of sensory stimuli appears to be beneficial when applied at the soles during spaceflight (Layne et al. 1998) and, plausibly applicable to a deconditioned astronaut, terrestrial compression garments can improve somatosensation in those with poorer somatosensation (Broatch et al. 2021) and improve aspects of balance control or in poorer performers (e.g., Noé, Baige and Paillard 2022).

Axial-loading countermeasure suits that have been used in space include the Russian Penguin (Pingvin) Suit (Kozlovskaya, Grigoriev and Stepantzov 1995, Yarmanova et al. 2015) and the gravity loading countermeasure skinsuit (GLCS; Waldie and Newman 2011, Bellisle and Newman 2020, trialled by ESA in 2015 and 2017). The Penguin suit uses elastic bungee cords that span from the shoulders to feet with an inelastic waist belt acting as a load divider; there is also the option of using stirrup bungee cords to provide passive stretching of the calf muscles and resistance to plantarflexion (Kozlovskaya, Grigoriev and Stepantzov 1995, Yarmanova et al. 2015).

Instead of bungee cords, the GLCS was designed as a skinsuit, using the suit fabric itself to incrementally load the torso and lower limbs in the axial direction via incremental suit strain, while using circumferential fabric strain (i.e., general

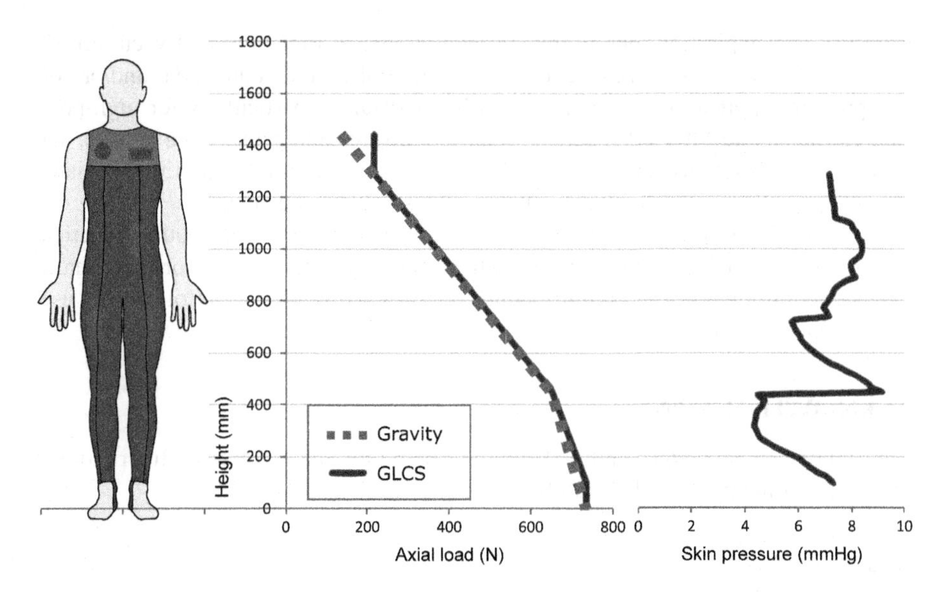

FIGURE 12.3 Gravity loading countermeasure skinsuit (GLCS). The axial load (here equivalent to Earth axial load; i.e., 1 *g*) is for an example person with a stature of 1.7 m and mass of 75 kg. The skin pressure profile on the far right is the calculated estimates of the local suit pressure required to anchor the suit against the skin to support the progressive axial loading. (Redrawn from Waldie and Newman, 2011.)

body compression) to anchor the axial load (Waldie and Newman 2011). In this way, the suit more closely resembles the way in which body weight increases incrementally when upright under the influence of gravity as the cumulative body mass above any given site increases (Figure 12.3). Considerable work has focused on improving and assessing the GLCS, often in the context of musculoskeletal targets (Bellisle and Newman 2020), while subsequent work has shifted emphasis more specifically toward neuromuscular and sensorimotor functionality (Bellisle et al. 2022, MacRae et al. 2022). This shift in focus is in part aligning with the need for countermeasures to mitigate postflight postural and locomotor control (Mulavara et al. 2018, Macaulay et al. 2021) and the likely importance of dynamic loading for bone preservation and formation (Turner 2004), which may not be sufficiently provided with axial-loading suits.

12.4 OUTLOOK

Compression textiles have existing applications in the context of spaceflight and may have a greater role to play in the future with a return to extraterrestrial surface exploration. Garments or suits using compression textiles inherently rely on the relationship between the dimensions of the suit and the dimensions of the underlying body part. For this reason, common areas of importance include: (1) understanding fabric stress–strain properties and changes with use (e.g., creep, stress relaxation, degradation with exposure to perspiration, environmental contaminants, radiation, and vacuum); (2) methods of adjustment and associated verification of correct adjustment and/or correct pressure application; (3) donning/doffing assistance (particularly for high-pressure garments like MCP EVA suits); (4) custom fitting (sizing and manufacture); and (5) understanding body anthropometric changes with spaceflight and the associated effects on loading or pressure application and associated adjustment.

Further, it is important to remain cognizant of operational and other practical considerations, such as compatibility with other equipment, cleaning or other maintenance requirements, toilet requirements, duration to don/doff, comfort during use, and suit/garment volume and mass.

ACKNOWLEDGMENT

The authors acknowledge funding from the Australian Space Agency's International Space Investment: Expand Capability Grant.

NOTE

1. From an engineering perspective the lower limit of space is ~80 km (McDowell 2018), while many organizations consider the boundary of space to be the so-called Kármán line at 100 km. That said, even these thresholds still actually fall within layers of Earth's atmosphere (mesosphere and thermosphere for the 80 km and 100 km thresholds, respectively).

REFERENCES

Abercromby, A F J, Bekdash, O, Cupples, J S, Dunn, J T, Dillon, E L, Garbino, A, Hernandez, Y, et al. 2020. "Crew health and performance extravehicular activity roadmap: 2020. NASA/TP-20205007604."

Annis, J F, and Webb, P. 1971. "Development of a space activity suit. NASA CR-1892." Washington, DC.

Bellisle, R F, and Newman, D. 2020. "Countermeasure suits for spaceflight." 50th International Conference on Environmental Systems 2020–315: 1–12.

Bellisle, R F, Ortiz, C, Porter, A, Harvey, A, Arquilla, K, Bjune, C, Waldie, J, and Newman, D. 2022. "The Mk-7 gravity loading countermeasure skinsuit: Evaluation and preliminary results." In 2022 IEEE Aerospace Conference (AERO), 1–11. IEEE. doi: 10.1109/aero53065.2022.9843474.

Bishop, P A, Lee, S M C, Conza, N E, Clapp, L L, Moore, A D, Williams, W J, Guilliams, M E, and Greenisen, M C. 1999. "Carbon dioxide accumulation, walking performance, and metabolic cost in the NASA launch and entry suit." *Aviation, Space, and Environmental Medicine* 70 (7): 656–665.

Broatch, J R, Halson, S L, Panchuk, D, Bishop, D J, and Waddington, G. 2021. "Compression enhances lower-limb somatosensation in individuals with poor somatosensation, but impairs performance in individuals with good somatosensation." *Translational Sports Medicine* 4: 280–288. doi: 10.1002/tsm2.214.

Buckey, J C, Lane, L D, Levine, B D, Watenpaugh, D E, Wright, S J, Moore, W E, Gaffney, F A, and Blomqvist, C G. 1996. "Orthostatic intolerance after spaceflight." *Journal of Applied Physiology* 81 (1): 7–18.

Bunford, R, MacRae, B, Waldie, J, Cable, G, and Padhye, R. 2022. "Orthostatic intolerance garments for spaceflight: Posture-informed design for improving garment comfort." 51st International Conference on Environmental Systems 2022–151: 1–10.

Carvil, P A T. 2017. "Axial loading as a countermeasure to microgravity-induced deconditioning; effects on the spine and its associated structures." King's College London.

Chappell, S P, Norcross, J R, Abercromby, A F, Bekdash, O S, Benson, E A, Jarvis, S L, Conkin, J, et al. 2017. "Risk of injury and compromised performance due to eva operations. hrp evidence report." Washington.

Clément, G. 2011. *Fundamentals of Space Medicine*. 2nd ed. New York: Springer. doi: 10.1007/1-4020-3434-2.

Crawford, I. 2012. "Dispelling the myth of robotic efficiency." *Astronomy and Geophysics* 53 (2): 22–26.

Denq, J C, Opfer-Gehrking, T L, Giuliani, M, Felten, J, Convertino, V A, and Low, P A. 1997. "Efficacy of compression of different capacitance beds in the amelioration of orthostatic hypotension." *Clinical Autonomic Research* 7 (1997): 321–326.

Diedrich, A, and Biaggioni, I. 2004. "Segmental orthostatic fluid shifts." *Clinical Autonomic Research* 14: 146–147. doi: 10.1007/s10286-004-0188-9.

Drake, B G. 2009. "Human exploration of mars design reference architecture 5.0 addendum. NASA special publication 2009–566-ADD." Hanover, MD. https://www.nasa.gov/pdf/373667main_NASA-SP-2009-566-ADD.pdf.

Granberry, R M, Eschen, K P, Ross, A J, Abel, J M, and Holschuh, B T. 2020. "Dynamic countermeasure fabrics for post-spaceflight orthostatic intolerance." *Aerospace Medicine and Human Performance* 91 (6): 525–531.

Green, D A, and Scott, J P R. 2018. "Spinal health during unloading and reloading associated with spaceflight." *Frontiers in Physiology* 8 (1126): 1–7. doi: 10.3389/fphys.2017.01126.

Holschuh, B T, and Newman, D J. 2016. "Morphing compression garments for space medicine and extravehicular activity using active materials." *Aerospace Medicine and Human Performance* 87 (2): 84–92.

Huerta, R, Kerr, E S, and Anderson, A P. 2018. "Mechanical counterpressure and gas-pressurized fusion spacesuit concept to enable martian planetary exploration." 48th International Conference on Environmental Systems 326 (2018): 1–15.

Jenkins, D R. 2012. "Dressing for altitude: U.S. Aviation pressure suits – Wiley post to space shuttle." NASA SP-2011-595.

Jessup, J M. 2015. "Hybrid enhanced epidermal spacesuit design approaches." University of North Dakota.

Kent, M. 2007. "Reference man and woman." The Oxford Dictionary of Sports Science & Medicine (Online Version). https://www.oxfordreference.com/view/10.1093/acref/9780198568506.001.0001/acref-9780198568506-e-5880.

Kluis, L, and Diaz-Artiles, A. 2021. "Revisiting decompression sickness risk and mobility in the context of the smartsuit, a hybrid planetary spacesuit." *NPJ Microgravity* 7 (46): 1–8. doi: 10.1038/s41526-021-00175-3.

Korth, D W. 2015. "Exercise countermeasure hardware evolution on ISS: The first decade." *Aerospace Medicine and Human Performance* 86 (12): A7–13. doi: 10.3357/AMHP.EC02.2015.

Kozlovskaya, I B, Grigoriev, A I, and Stepantzov, V I. 1995. "Countermeasure of the negative effects of weightlessness on physical systems in long-term space flights." *Acta Astronautica* 36 (8–12): 661–668. doi: 10.1016/0094-5765(95)00156-5.

Kozlovskaya, I B, Yarmanova, E N, and Fomina, E V. 2015. "The Russian system of preventive countermeasures: Its present and future." *Human Physiology* 41 (1): 704–711. doi: 10.1134/S0362119715070075.

Kracik, M, Meyen, F, Trotti, G, and Newman, D J. 2012. "The development of a high mobility space suit helmet for planetary exploration." In 63rd International Astronautical Congress.

Layne, C S, Mulavara, A P, Pruett, C J, McDonald, P V, Kozlovskaya, I B, and Bloomberg, J J. 1998. "The use of in-flight foot pressure as a countermeasure to neuromuscular degradation." *Acta Astronautica* 42: 231–246.

Lee, S M C, Guined, J R, Brown, A K, Stenger, M B, and Platts, S H. 2011. "Metabolic consequences of garments worn to protect against post-spaceflight orthostatic intolerance." *Aviation, Space, and Environmental Medicine* 82 (6): 648–653. doi: 10.3357/ASEM.3039.2011.

Lee, S M C, Ribeiro, L C, Laurie, S S, Feiveson, A H, Kitov, V V, Kofman, I S, Macias, B R, et al. 2020. "Efficacy of gradient compression garments in the hours after long-duration spaceflight." *Frontiers in Physiology* 11 (784): 1–10. doi: 10.3389/fphys.2020.00784.

Macaulay, T R, Peters, B T, Wood, S J, Clément, G R, Oddsson, L, and Bloomberg, J J. 2021. "Developing proprioceptive countermeasures to mitigate postural and locomotor control deficits after long-duration spaceflight." *Frontiers in Systems Neuroscience* 15 (658985): 1–14. doi: 10.3389/fnsys.2021.658985.

MacRae, B A, Bunford, R, Waldie, J, Rudakov, A, and Padhye, R. 2022. "Development of a mechanical-loading countermeasure skinsuit to mitigate post-spaceflight sensorimotor dysfunction." 51st International Conference on Environmental Systems 82: 1–12.

MacRae, B A, Laing, R M, and Partsch, H. 2016. "General considerations for compression garments in sports: Applied pressures and body coverage." In *Compression Garments in Sports: Athletic Performance and Recovery*, 1–32. doi: 10.1007/978-3-319-39480-0_1.

McDowell, J C. 2018. "The edge of space: Revisiting the karman line." *Acta Astronautica* 151: 668–677. doi: 10.1016/j.actaastro.2018.07.003.

McFarland, S M. 2022. "Mechanical counter-pressure EVA suits: Nasa outlook and development strategy in 2022." 51st International Conference on Environmental Systems 2022–249: 1–14.

McFarland, S M, Ross, A J, and Saunders, R W. 2019. "The 'Space activity suit' – A historical perspective and a primer on the physiology of mechanical counter-pressure." 49th International Conference on Environmental Systems 173 (2019): 1–30.

Mulavara, A P, Peters, B T, Miller, C A, Kofman, I S, Reschke, M F, Taylor, L C, Lawrence, E L, et al. 2018. "Physiological and functional alterations after spaceflight and bed rest." *Medicine and Science in Sports and Exercise* 50 (9): 1961–1980. doi: 10.1249/MSS.0000000000001615.

NASA Office of Inspector General. 2021. "NASA's development of next-generation spacesuits. Report no. IG-21-025." Washington. https://oig.nasa.gov/docs/IG-21-025.pdf.

Newman, D J, Canina, M, and Trotti, G L. 2007. "Revolutionary design for astronaut exploration - beyond the biosuit system." American Institute of Physics Conference Proceeding 880: 975–986. doi: /10.1063/1.2437541.

Nicogossian, A E, Williams, R S, Huntoon, C L, Doarn, C R, Polk, J D, and Schneider, V S. 2016. *Space Physiology and Medicine; From Evidence to Practice* (Fourth Edition). New York: Springer.

Noé, F, Baige, K, and Paillard, T. 2022. "Can compression garments reduce inter-limb balance asymmetries?" *Frontiers in Human Neuroscience* 16 (835784): 1–9. doi: 10.3389/fnhum.2022.835784.

Platts, S H, Tuxhorn, J A, Ribeiro, L C, Stenger, M B, Lee, S M C, and Meck, J V. 2009. "Compression garments as countermeasures to orthostatic intolerance." *Aviation, Space, and Environmental Medicine* 80 (5): 437–442. doi: 10.3357/ASEM.2473.2009.

Rowell, L B, Detry, J-M R, Blackmon, J R, and Wyss, C. 1972. "Importance of the splanchnic vascular bed in human blood pressure regulation." *Journal of Applied Physiology* 32 (2): 213–220.

Rudakov, A, Clarke, J, MacRae, B A, Waldie, J, and Padhye, R. 2022. "Feasibility of MCP and hybrid GP/MCP architectures for martian EVA: A trade study perspective." 51st International Conference on Environmental Systems 2022–81: 1–15.

Schauss, G, Bellisle, R, Kothakonda, A, Newman, D, and Anderson, A. 2022. "High performance mechanical counter-pressure spacesuit glove for martian surface exploration." 51st International Conference on Environmental Systems 191 (2022): 1–12.

Shigueva, T A, Zakirova, A Z, Tomilovskaya, E S, and Kozlovskaya, I B. 2015. "The influence of support and weight unloading on the characteristics of spinal reflex." In IAA 2015 Humans in Space Symposium, 171. Prague: International Academy of Astronautics.

Southern, T C, and Moiseev, N A. 2017. "Novel mechanical counter pressure gloves." 47th International Conference on Environmental Systems 2017–97: 1–15.

Stenger, M B, Brown, A K, Lee, S M C, Locke, J P, and Platts, S H. 2010. "Gradient compression garments as a countermeasure to post-spaceflight orthostatic intolerance." *Aviation, Space, and Environmental Medicine* 81 (9): 883–887. doi: 10.3357/ASEM.2781.2010.

Stenger, M B, Lee, S M C, Ribeiro, L C, Phillips, T R, Ploutz-Snyder, R J, Willig, M C, Westby, C M, and Platts, S H. 2014. "Gradient compression garments protect against orthostatic intolerance during recovery from bed rest." *European Journal of Applied Physiology* 114: 597–608. doi: 10.1007/s00421-013-2787-4.

Stenger, M B, Lee, S M C, Westby, C M, Ribeiro, L C, Phillips, T R, Martin, D S, and Platts, S H. 2013. "Abdomen-High elastic gradient compression garments during post-spaceflight stand tests." *Aviation, Space, and Environmental Medicine* 84 (5): 459–466. doi: 10.3357/ASEM.3528.2013.

Stroming, J, and Newman, D J. 2020a. "Design of an external compensatory breathing bladder for the biosuit." In 2020 IEEE Aerospace Conference, 1–9.

Stroming, J, and Newman, D J. 2020b. "Thermal modeling of mechanical counterpressure spacesuit EVA." International Conference on Environmental Systems 2020–234: 1–12.

Sullivan, K D. 2019. *Handprints on Hubble: An Astronaut's Story of Invention*. Cambridge: The MIT Press.

Thomas, K S, and McMann, H J. 2012. *U. S. Spacesuits* (2nd edition). New York: Springer-Verlag.

Turner, C H. 2004. "Biomechanical aspects of bone formation." In *Bone Formation*, edited by F Bronner and M C Farach-Carson, 79–105. London: Springer-Verlag. doi: 10.1533/9781845696610.1.106.

Vil-Viliams, I F, Kotovskaya, A R, Gavrilova, L N, Lukjanuk, V Y, and Yarov, A S. 1998. "Human +Gx tolerance with the use of Anti-G suits during descent from orbit of the soyuz space vehicles." *Journal of Gravitational Physiology* 5 (1): 129–130.

Waldie, J M. 2005. "Mechanical counter pressure space suits: Advantages, limitations and concepts for martian exploration." *The Mars Society*: 1–17. http://www.marspapers.org/#/papers.

Waldie, J M, and Newman, D J. 2011. "A gravity loading countermeasure skinsuit." *Acta Astronautica* 68 (7–8): 722–730. doi: 10.1016/j.actaastro.2010.07.022.

Waldie, J M, Tanaka, K, Tourbier, D, Webb, P, Jarvis, C W, and Hargens, A R. 2002. "Compression under a mechanical counterpressure space suit glove." *Journal of Gravitational Physiology* 9 (2): 93–97.

Webb, P, and Annis, J F. 1967. "The principle of the space activity suit. NASA CR-973." Washington, DC.

Wickman, L A, and Luna, B. 1996. "Locomotion while load-carrying in reduced gravities." *Aviation, Space, and Environmental Medicine* 67 (10): 940–946.

Yamashita-Goto, K, Okuyama, R, Honda, M, Kawasaki, K, Fujita, K, Yamada, T, Nonaka, I, et al. 2001. "Maximal and submaximal forces of slow fibers in human soleus after bed rest." *Journal of Applied Physiology* 91 (2001): 417–424.

Yarmanova, E N, Kozlovskaya, I B, Khimoroda, N N, and Fomina, E V. 2015. "Evolution of russian microgravity countermeasures." *Aerospace Medicine and Human Performance* 86 (12): A32–A37. doi: 10.3357/AMHP.EC05.2015.

Zhang, S, Wang, K, Zhu, R, Jiang, C, and Niu, W. 2021. "Penguin suit and fetal position finite element model to prevent low back pain in spaceflight." *Aerospace Medicine and Human Performance* 92 (5): 312–318.

Index

Printed in the United States
by Baker & Taylor Publisher Services

Printed in the United States
by Baker & Taylor Publisher Services